BORDERS IN SERVICE

Enactments of Nationhood in Transnational Call Centres

Edited by Kiran Mirchandani and Winifred R. Poster

Borders in Service traces the intersection of service labour and national identity across global call centres in El Salvador, Guatemala, Guyana, Mauritius, Morocco, and the Philippines and at the U.S.-Mexico border. While most studies of offshore call centres have focused on India, this collection of essays examines the experiences of call centre workers in many of the newly emerging hubs of transnational service work.

In this volume, editors Kiran Mirchandani and Winifred R. Poster have gathered an international group of contributors from a variety of disciplines to explore many of the issues affecting work within global call centres, such as language, accent, and conversational dynamics; education and training; physical space; and organizational, human resource, and labour practices and policies. By grounding the theoretical debates on nationhood and labour in the realities of daily life in global call centres, *Borders in Service* offers a timely, accessible, and revealing study that will change our thinking about outsourced customer service work.

KIRAN MIRCHANDANI is a professor in the Department of Leadership, Higher and Adult Education at the Ontario Institute for Studies in Education, University of Toronto.

WINIFRED R. POSTER is an instructor at Washington University in St Louis.

BORDERS IN SERVICE

Enactments of Nationhood in
Transnational Call Centres

Edited by Kiran Mirchandani
and Winifred R. Poster

UNIVERSITY OF TORONTO PRESS
Toronto Buffalo London

© University of Toronto Press 2016
Toronto Buffalo London
www.utppublishing.com

ISBN 978-1-4875-0080-1 (cloth)   ISBN 978-1-4875-2059-5 (paper)

Library and Archives Canada Cataloguing in Publication

Borders in service : enactments of nationhood in transnational
call centres/edited by Kiran Mirchandani and Winifred R. Poster.

Includes bibliographical references.
ISBN 978-1-4875-0080-1 (cloth).–ISBN 978-1-4875-2059-5 (paper)

1. Call Centers–Social aspects.   2. Call centers–Employees.
3. Service industries workers.   4. Intercultural communication.
5. National characteristics.   6. Transnationalism.   I. Mirchandani, Kiran,
1968–, author, editor   II. Poster, Winifred, author, editor

HE8788.B67 2016   381   C2016-903988-9

University of Toronto Press acknowledges the financial assistance to its
publishing program of the Canada Council for the Arts and the Ontario
Arts Council, an agency of the Government of Ontario.

Canada Council **Canada Council** **Conseil des Arts**
for the Arts **for the Arts** **du Canada**

ONTARIO ARTS COUNCIL
CONSEIL DES ARTS DE L'ONTARIO
an Ontario government agency
un organisme du gouvernement de l'Ontario

Funded by the   Financé par le
Government   gouvernement
of Canada   du Canada

Canadä

Contents

Illustrations

Acknowledgments

We are grateful to the University of Toronto's Visiting Speaker Program at the Ontario Institute for Studies in Education, which facilitated the collaboration that led to this edited collection. We had the benefit of working with some excellent graduate students in developing and deepening our understanding of the themes, and four in particular – Meaghan Brugha, Hoda Farahmandpour, Rozalina Omar, and Dulani Suraweera – provided outstanding research assistance. The staff at the University of Toronto Press was a pleasure to work with and provided timely and valuable feedback throughout. The anonymous reviewers were incredibly generous in their thoughtful and constructive responses. This project was supported by the Social Sciences and Humanities Research Council of Canada.

Working with authors from around the world has been an exciting and satisfying experience, and we are deeply grateful for the enthusiasm of our contributors. It has been a privilege for us, as editors, to discover the overlapping and divergent practices and processes that have been revealed through their original field research. We have had the opportunity to engage in virtual ethnographies through our discussions with authors about the fascinating configurations of call centre work in different countries. Fruitful conversations were exchanged at sessions on this theme that we organized at the International Labour Process Conference in Berlin in 2016.

Finally, we are grateful to each of our local intellectual communities, our friends, and our families for providing the bedrock that is foundational to the success of all our projects. Kiran would like to thank Enakshi Dua, Yasmin Gopal, Ashwin, Suvan and Syona Joshi, Audrey Macklin,

Ajit, Sheeley, Sandeep, Ria and Kai Mirchandani, Shahrzad Mojab, Pushkala Prasad, Alissa Trotz, Leah Vosko, and Jingjing Xu. Winnie appreciates the scholarly advice of Miriam Cherry, the intellectual spirit of the Labor Tech Reading Group, and the ongoing support of Sanjay Jain, Ketan Jain-Poster, Natasha Jain-Poster, Jamie Poster, Stacey Herzing, Marianne Ross, Helena Ross, Claire Poster, June Poster, Joy Karol, Annette Schlichter, and in India the Jain family: PK, Meera, Vikas, Shikha, Vidhi, Pranav, Mohit, Seeru, Rohan, Madhvi, Mukesh, Rachita, Bhavya, and Siddhant.

Kiran Mirchandani and Winifred R. Poster

Introduction

1 Enactments of Nationhood in Transnational Call Centres

WINIFRED R. POSTER AND KIRAN MIRCHANDANI

The aim of *Borders in Service* is to trace the connections between two dynamics that underlie transnational customer service encounters between employees and customers – service labour and constructions of nationhood. We argue that the transnational service industry, of which the call centre is an important site, allows for a cogent analysis of these two previously unconnected dynamics of labour and nation. While *labour* has been explored largely in relation to notions of economy and market, and *nation* in relation to politics and the state, this collection draws attention to their intersection: the labour involved in constructing nations, and the nationalisms implicit (and at times explicit) in doing service labour.

Given the fundamentally embodied nature of service work, in which the body or voice of the worker is an intrinsic part of the commodity being exchanged, the identities, social locations, and national contexts of workers and customers form the unspoken bedrock upon which service work is conducted and evaluated. This book explores the work of people engaged in transnational interactive service work who, as part of their jobs, give "concrete expression to their understanding of the nation" (Fox and Miller-Idriss, 2008, p. 539). The contributors to this collection are researchers who are located in a wide array of national settings and have conducted original ethnographic field research on call centres. They explore the political economy of nation in the site of the global call centre – that is, the ways in which borders are enacted, and ideals of nationhood and citizenship are created and expressed through encounters between providers and recipients during transnational voice-based service interactions. Collectively, the chapters in this book capture the diverse ways in which nationalisms are enacted through the everyday interactions between transnational call centre workers, their employers, and their customers.

Overview of Global Call Centres

Transnational Services and Communicative Capitalism

Call centres emerged as an interdisciplinary curiosity in the last decades of the twentieth century. A *call centre* is an organization, or part of an organization, that handles telephone communications with the public. These involve either the *inbound* labour of receiving customer queries about products, scheduling appointments, providing technical help, problem solving, etc., or the *outbound* labour of calling customers for telemarketing, debt collections, etc.

Many scholarly volumes and analyses have reflected on the dynamics of these *domestic* call centres, that is, those that are located in the same country as the client firms for which they sell and the consumer populations that they serve (for some insightful reviews and collections see Burgess & Connell, 2006; Deery & Kinnie, 2004). Call centres have become focal points of interest across diverse fields because they represent a convergence of many trends in the study of the sociology of work – labour control, resistance, identity construction, emotion work, professionalization, and gendering practices.

Starting around the year 2000, however, this industry transformed radically. Call centres became *global* with the rapid acceleration of outsourcing and offshoring. *Outsourcing* moves the production process out of the office, often to other specialized firms that perform customer service exclusively. *Offshoring* is the process of sending work across national borders and then retrieving the end products or services by mail, shipping, or electronically. In their drive for lower labour and infrastructure costs, firms displace work not only from the main employer (as in domestic outsourcing) but also from the customer base.

Global call centres, in turn, have arisen as the incarnation of this process; that is, firms are diverting their local calls through satellite connections and across national lines, and back to consumers, sometimes under the presumption that the call has never left the original firm or country. Distinctive about these organizations, moreover, is their crossing of lines between the global north and south. The global north is conceptualized as including wealthy countries like the United States, Canada, Russia, Japan, Australia, and many in Europe. The global south is said to include emerging economies in Central and South America, Asia, and Africa. This general (though not exclusive) mapping of affluence and political power along geographic poles is said to be significant for contemporary transnational relations. Hence, the spread of the global call centre industry across this map has important implications for the dynamics of labour, as we will discuss. With this recent switch of outsourcing destinations by firms comes a new constellation in which workers in the global south are largely making calls to and receiving calls from consumers in the global north.

The labour of global call centres should not be confused with that of other types of outsourcing. Rather, it represents a significant shift in the foundation and direction of capitalism in the post-industrial era. We argue that global call centres encapsulate a merging of two new trends in the global economy. One is an expansion of transnational services. A service industry involves doing things for people (rather than making things, as in factories). Service jobs pervade economic sectors – from government (for example, social workers) to transportation (for example, truck drivers), to health (for example, nurses), to education (for example, teachers), and many more. They also run up and down the ladder of occupational levels, from the very high-skilled workers (for example, finance, legal, and medical professionals) to the very low-skilled workers (for example, restaurant dishwashers and janitors). Indeed, services are now the fastest-growing jobs in formal sectors of the economy around the world (Poster & Yolmo, 2016). People refer to many service jobs as front-line work, meaning that employees are at the interface of firms and their consumers (as opposed to the behind-the-scenes labour of employees whom consumers never see, for instance in data entry and warehouse stocking).

The second relevant trend in post-industrial capitalism is the rise of communicative industries. This sector focuses entirely on exchanges between people, especially those exchanges that are mediated rather than face to face – that is, written on paper, posted on the Internet, sent via video, or, in our case, spoken on the telephone. In what Dean (2009) calls "communicative capitalism," this industry facilitates and/or profits from the creation of communications between citizens. At times, it *directly produces* the communicative infrastructure (for example, telecommunications corporations that sell phone services; electronics factories that produce mobile phones; internet and cable providers; social networking companies, etc.). At other times it *indirectly profits* from communications (for example, marketing companies that analyse the content of posts, texts, and tweets).

Less studied, however, is yet another crucial element of the communicative economy: the business of facilitating exchanges between firms and their customers. Brophy (2010) places call centres at the nexus of this dynamic because they manage customer service relations for all kinds of organizations, public and private.

This merging of the service and communication industries changes the types of labour that are sent overseas within outsourcing and offshoring. Rather than sending only manufacturing jobs, firms are increasingly moving abroad the work of information, data, and phones. Early periods of outsourcing in the 1980s and 1990s initiated this process with information services: high-end software development and low-end data entry. Emblematic of the twenty-first century, however, is the nascence of transnational communicative industries: mid-range labour in phone-based customer service. Global south states and firms, in turn, are selling an embodied form of communication. As Cecilia Rivas (chapter 2) recounts, states and their firms are advertising how they "export voice."

In many countries around the world, global call centres comprise a significant sector of the economy, with some, such as those in India, serving 73 per cent foreign clients (Holman et al., 2007). Four million Americans and 800,000 people in the United Kingdom work in call centres. In Australia 75 per cent of customer contact occurs via call centres (Russell 2009, p. 6). As Glucksmann has observed, "call centres represent not only the most rapidly expanding forms of work and of business organization but also one of the most researched" (2004, p. 795).

Attention towards these global call centres has prompted a surge of literature (Russell 2008; Batt et al., 2009). Studies have shed light on the labour processes that are common across diverse national locations, such as hyper-routinization, scripting, surveillance, temporal controls, and emotional stress. Some reflect on the widespread recognition of the importance of national economic policies in facilitating trade in services, especially those from divergent contexts (Huws 2009). Yet, little is known about how the nation itself is conceptualized and constructed via and within transnational call centres.

India and Beyond: Jumping Off from the Early Case Studies

India has been the focal point for a majority of the studies on transnational call centres in the first decade and a half of their emergence (Aneesh, 2006, 2015; Basi, 2009; Das et al., 2008; Nadeem, 2011; Noronha and D'Cruz, 2009; Patel, 2010; Rowe et al., 2013; Russell, 2009; Taylor and Bain, 2005; Thite and Russell, 2009). The rapid growth of the industry in India is likely part of the reason that this site has been identified as important by researchers working in diverse disciplines and contexts.

Indeed, India was among the first to build and receive the telecommunications infrastructure needed for this transnational industry. Panicked about the "Y2K" (year 2000) computer virus, U.S. and Indian entrepreneurs pushed for the laying of fibre optic cables underneath the Atlantic Ocean so that Indian engineers could develop and send software patches digitally to the United States. Other technologies and economies also converged at this time: expanded access to satellite communications systems, a significant reduction in the price of Internet broadband, and the development of programs such as VOIP (voice over Internet protocol), which enables phone calls through the Internet. Indian entrepreneurs, meanwhile, had forged pivotal ties with business contacts in the United States, Canada, and the United Kingdom through previous decades of immigration.

Scholarship on this initial case of Indian call centres, then, has cued us into critical dynamics of outsourcing. Substantively, it reveals the key role of nationality and nationhood within these transnational labour processes. A provocative thread running through the work of these researchers, including the editors of this volume, has been the constructions of nation through identity, culture, and language. Poster's (2007) work, for example, reveals a process of "national identity management" in which managers and clients of outsourced call centres place particular requirements on workers. Beyond expectations of displaying pleasantness and subduing anger, they also ask workers to display "Americanness." So, in the process of selling mortgage insurance or helping to solve computer problems, call centre workers are encouraged to use American names, adopt American accents, convey Americanness through idle conversation, and use a range of prepared scripts to state implicitly or explicitly that they are in the United States. These four practices fall on a continuum of styles of locational masking, ranging from the lesser forms that are indirect and more suggestive, to the more extreme forms that involve direct and outright lying. Employers vary in how extensively they are committed to this endeavour, and the lengths they will go to promote the facade. In some call centres, workers can be fired for failing to carry out various elements of this process effectively.

Mirchandani (2012) explores the frequency of customer anger during customer service encounters and the consequences of this customer

anger for nation building in the West. While angry customers exacer-
bate worker stress and intensify the need for training, they also serve
a "productive" purpose in the everyday creation and perpetuation of
Western nationalism. She argues that Western state and public discourses on offshoring sanction customer aggression on calls; these calls
provide opportunities for customers to exercise citizenship rights over
jobs that are assumed to have been stolen. In so doing, customers in the
West continually define and then protect the assets (jobs) that belong to
the nation. Thus, they enact patriotism through their anger.

These insights are especially important given that, over the past five
years, there has been significant dispersal of the transnational call centre industry. Although estimates are crude, some analysts predict that
700,000 jobs shifted from India to other countries between 2008 and 2013.
These new destinations include Canada, Costa Rica, Mexico, Brazil, the
Czech Republic, Egypt, Bulgaria, Romania, Poland, the Ukraine, China,
the Philippines, Australia. and Malaysia. Large Indian companies have
also established a significant presence as call centre providers in numerous countries, such as the Philippines (Arun, 2013). Reports indicate that
there are now one million transnational call centre workers in the Philippines (Lee, 2015). Countries such as Kenya, Rwanda, and South Africa
have emerging call centre sectors that serve English-speaking customers
(Benner, 2006; Free, 2014; Graham & Mann, 2013; Hunter & Hachimi,
2012; Mann, Graham, & Friederici, 2014). French-language call centres
are opening in Morocco, Senegal, Tunisia, and Madagascar (Lacey, 2005).
The global call centre industry now facilitates the crossing of multiple
national borders during daily service interactions.

However, this kind of emphasis on nationhood and nationalism has
been less prevalent in the research on global call centres in other parts of
the world. As an elaboration of our approach, this collection extends
and deepens the analysis by exploring connections between labour and
nation in the context of proliferating transnational call centres and in a
variety of other cross-national interactions.

Voice and Talk: The Labour of Communicative Interactive Service

Talking on the phone is an unusual job. As a task it may be a common feature of contemporary work, but as an occupation in itself it
has specific requirements for the service industry. Moreover, outsourcing changes the substance of what people talk about by placing those
conversations in international contexts. Several dynamics, then, make

nation and nationalism especially salient within call centre settings, and thus unique from other kinds of transnational services.

First, unlike other kinds of service labour, call centre work is *interactive*. Employees are in direct contact with customers, clients, and colleagues (at least through verbal exchanges). Moreover, transnational call centres bring service participants from different countries in voice-to-voice contact on a mass scale, perhaps for the first time. In turn, these actors carry the nationally-grounded social meanings and habits of their everyday lives with them in their conversations on the telephone.

Second, customer service invokes the symbols and weight of nation within the process of *consumption*. A main purpose of call centres is to facilitate the exchange of goods with the public. Yet those goods (that is, the brands) are imbued with nationhood, and accordingly so will be the substance of the interactions about those goods on the telephone. Moreover, the service process may be viewed through a national and global lens. As revealed by the concept of *service ethnocentrism*, consumers often express preferences about the nationalities of the workers who are providing service (Thelen, Honeycutt Jr., & Murphy, 2010; Thelen, Thelen, Magnini, & Honeycutt Jr, 2009; Thelen, Yoo, & Magnini, 2010).

Customers evaluate the services they receive on the basis of pre-existing nationalist sentiments. A study of Americans' attitudes towards offshoring finds that while one would expect customers who have professionally benefited from offshoring to be positively predisposed towards the trend, in fact those "who feel a sense of national superiority, or who feel that members of other ethnic and racial groups are less praiseworthy than their own racial or ethnic group, tend to have particularly hostile reactions to outsourcing" (Mutz & Mansfield, 2011, p. 3), irrespective of their employment history.

Third, customer service is *communicative*. Quite critically, language has a special role in service, as the medium by which it is relayed. Accordingly, national features of speech – accent, voice, word choice, etc. – become key ingredients of the exchange. In global call centres we see a process of submerging, heightening, and altering national features of language. For instance, the same baseline language (in this case, English) becomes territorialized (as Indian English, Philippine English, etc.) and then re-territorialized (as globally neutral English). As Sonntag (2005) argues, global English is a form of "linguistic capital" that facilitates the need for Western corporations to acquire lower costs through offshoring, while it acts as a mechanism for cultural imperialism.

Fourth, *emotions* are integral to service transactions, and these may be imbued with nationalism. Research has shown how displays of emotion by employees, as well as expectations by consumers about the emotions of employees, are grounded in national foundations (Grandey, Fisk, & Steiner, 2005; Grandey, Rafaeli, Ravid, & Wirtz, 2010).

Fifth, customer service involves *aesthetic and bodily labour* (Lan, 2001; Wacquant, 1995), as workers' bodies are marked with nationhood. The interactive nature of call centre work, in conjunction with the important role played by the customer in the service encounter, reveals that the work of employees involves aesthetic labour (Warhurst & Nickson, 2007). Aesthetic labour refers to "the mobilization, development and commodification of embodied 'dispositions'" of the worker. These dispositions include his or her physique, bodily appearance, and even voice. Employers transform dispositions "into 'skills' ... geared towards producing a 'style' of service encounter that appeals to the senses of the customer" (Witz, Warhurst, & Nickson, 2003, p. 37).

Put simply, the aesthetic labour of call centres involves the work of sounding right over the telephone. The local dynamics of aesthetic labour have been well documented in other types of services (such as retail service and tourism), yet less attention has been devoted to how this is happening globally in sites like call centres. We will examine how employees often display subtle or overt signals of nationalism through their physical display – their clothing, appearance, voice, etc. – even though their work is on the telephone.

Finally, but not insignificantly, call centre work takes place entirely through technological platforms. The virtual context of communicative labour, in turn, shapes the way individuals present themselves to each other, as the medium of exchange is sound (audio) rather than sight (video or text) (Poster, 2015). Some analysts argue that this context heightens tendencies for consumers to express hostility and verbal abuse, especially related to xenophobia (D'Cruz & Noronha, 2014).

Moreover, this context enables capabilities for anonymizing and morphing. Speakers can alter their identities, presenting themselves with different nationalities to their callers. Knowing this, some firms are engaging in strategies of vocal manipulation among workers to prevent, or at least obscure, the revealing of their corporate identities vis-à-vis customers (Poster, 2015). In short, voice has utility for deceptability in this industry; it can help to maintain the anonymity of the firm as an outsourcer. In this project we explore the way that firms use the deceptive capacities of telephone work to manipulate nationality.

Thus, telephone work is inseparable from the local, embodied, situated contexts of workers. As a result, service offshoring illuminates many central components of the globalization of this work – such as identity, surveillance, control, culture, stratification – and nationalism. The book maps the diverse notions of nationhood that occur through these service interactions.

Key Concepts of Border Enactment in Transnational Call Centres

The National within the Transnational

This volume brings two new directions to the theorizing of the national and transnational in global call centres. First, we emphasize the *interplay* of the global and the national in terms of their unique coexistence and mutual constitution in the contemporary era. In many analyses of call centres, the global often takes centre stage. Considerable attention is given to features of the supranational (for example, neoliberalism, westernization, transnational flows, global divisions of labour, etc.). Some argue that the relevance of the nation is deteriorating in light of global trends. Thomas Friedman (2005), for instance, uses outsourced call centres as an example of a "flattening world" in which the trends of connectivity, workflow software, and other technologies are enabling "a global community that is soon going to be able to participate in all sorts of discovery and innovation" (p. 212), and "without regard to geography, distance, time, and, in the near future, even language" (pp. 205). Within the narrative of the "flat world," call centres are conceptualized as uplifting global south workers into the middle class as well as enhancing the economies of the global north through the expanding purchasing power of global south consumers.

Our argument concerns the continued relevance and in fact reification of nation in the transnational service economy. Indeed, Sassen (2006, p. 2) reminds us that "the epochal transformation we call globalization is taking place inside the national to a far larger extent than is usually recognized. It is here that the most complex meanings of the global are being constituted, and the national is also often one of the key enablers and enactors of the emergent global scale." Along these lines, the chapters in this collection reveal that *the production of nation by call centre actors is what enables the transnationalism of outsourcing to be carried out.* By recreating (sometimes fictitious) national symmetries between onshore consumers and offshore workers through talk on the telephone, global call centres bolster the facade of local production. This, in turn, facilitates the silent continuation of outsourcing practices, in which northern firms move their contracts to overseas locations with cheaper operating costs.

Conceptually (if not also methodologically) the articles in this collection adopt a frame of "global ethnography" (Burawoy, Blum, George, Gille, & Thayer, 2000) in order to view the macro and the micro in one setting and to see how the global, national, and local are engaged interactively. In the process we incorporate an analysis of the central features of race, ethnicity, sexuality, class, etc., within those sites (Poster, 2002). Thus, we engage with debates within the field of transnational studies on the salience or dominance of the global over the national. The chapters highlight how imageries and practices of the national coincide with, and at times counteract, those of the global – in short, how the two are in continual interplay and tension.

As a second direction for our volume, we question the conceptualization of nation in call centre analysis. Much of the writing on transnational call centres to date assumes a fixed national frame: the nation is often conceptualized as an uncontested, clearly demarcated, physical land mass. Many see it as "a bounded space under the control of a group of people, with fixed boundaries, exclusive internal sovereignty, and equal external status," similar to the way in which territory is conceptualized (Elden, 2013, p. 18). The nation may have emerged historically under particular circumstances and with a particular role: it aided the consolidation power of merchants and bankers in medieval times, in their drive to assert independence from feudal lords, aristocratic nobility, and religious leaders and to lay the foundations of capitalism (Sassen, 2006). Later, by functioning as the administrative centre of imperialism, the nation helped those capitalists to enhance their wealth. However, the expression of the state as a concrete and unified entity (if such a thing ever existed) is being increasingly destabilized.

Indeed, many scholars agree (to varying degrees) that the nation itself is a social construct, especially as a product of the contextualized interests of actors at particular points in time. It is as much about symbolic imaginings, communities, beliefs, and ideologies as it about material geographies, organizations, and policies. Our volume is an extension of the considerable empirical and theoretical research on how nations are socially fabricated, sometimes in very contradictory ways, and reflective of particular sectors of society that are in power at the moment. We examine related concepts of *nationhood* (understandings of what constitutes the nation), *national identity* (individual subjectivities and personal associations with the state), and *nationalism* (allegiance to the nation through heightened political zeal and often collective movements) as extensions and integral processes of nation building.

Thus, while rich data exist on call centre labour processes, technological standardization, worker experiences, and offshoring practices, little is known about how the very idea of nation is "imagined" (Anderson, 1991) as part of the work of interacting across national contexts. Towards this end we direct attention to a particular dynamic of nation making: the process of *banal nationalism*. This refers to the way that nation is constructed through the everyday practices of, in our case, global call centres. As Fox and Miller-Idriss (2008) note, the nation "is not simply the product of macro-structural forces; it is simultaneously the practical accomplishment of ordinary people engaged in routine activities." Nations are "produced" through "mundane practices" (Billig, 1995) such as routine talk and interactions, the choices that people make, the symbols that are celebrated, and the consumption preferences that are expressed (Fox & Miller-Idriss, 2008, p. 537).

Call centre workers, employers, and customers engage continually in these activities, through the training programs they design and in which they participate, the scripts they develop and use, and the ways they construct their own identities. In this sense the sector provides a useful site for the "empirically grounded investigation of nation in everyday life" (Fox & Miller-Idriss, 2008, p. 539). Based on interviews with members of the ethnic majority in Britain, Skey (2011) documents the ways in which everyday talk about lives, work, and news stories is guided by a "national frame of reference" that is taken for granted. He argues that "it is through peoples' daily engagement with, and mutual recognition of, these everyday symbols, institutional arrangements, familiar places and social practices that having a national community ... makes sense" (p. 152, italics removed).

In short, the essays in this collection interrogate the tensions of nationhood, while exposing its significance in work settings. We examine the contexts in which nationhood is maintained, weakened, or reified and heightened. In some sites it appears as a monolithic category (especially in opposition to, and in confrontation with, categories of nationhood from other geographic locations). But at other times nationhood is destabilized, questioned, and integrated with other features of other nations or with the transnational.

Post-Colonialism in Action: Reconnections of Empire and Colony

Given the authority of neoliberalism, one might assume that global firms transcend concerns about nation in favour of profit. For instance, labour arbitrage is said to motivate global industries to move to locations in the global south where costs are cheapest. This is especially true of industries like electronics and garment manufacturing (Poster & Yolmo, 2016), where firms are known to establish themselves in countries that have the highest rates of poverty and the lowest wages (like Bangladesh and Thailand).

Yet, this is not necessarily the case with global call centres. Entrepreneurs in this industry have another agenda for their practice of international outsourcing: to find a communicative match between their own consumers and potential workforces abroad. This refers, in a direct sense, to language. Workers and consumers should speak the same language on the telephone, despite geographic distance, to ensure a baseline understanding of one another. In addition, the communicative match has a broader definition. It refers to sociocultural understandings about the nation, including points such as religion, government, economic policy, entertainment, and leisure. Workers and consumers should share parallel meanings of the wider national context within their conversations, whether formal or informal.

As a default strategy for achieving this symmetry, *global firms have been reconnecting former national pairings of empire and colony*. Contracts from the global north tend to go to locations in the global south where their representative nations have previously engaged in colonization (or where their colonized subjects have migrated): United States–Philippines, United Kingdom–India, and France–Algeria, for example. This is also the case in China, where workforces in different regions of the country are targeted by global firms depending on their prior histories of conquest and migration (Qiu, 2010): Mandarin-speaking populations in the southeastern provinces of Guangdong and Fujian are sought by firms from Taiwan; Cantonese-speaking populations in Guangzhou are sought by firms in Hong Kong; Japanese speakers in northeast Manchuria are sought by firms in Japan, owing to its historic occupation of that region.

In short, what we see in the global call centre industry is an overt system of post-colonial regroupings. Observing this trend, a new wave of scholarship is calling for the adoption of a post-colonial framework to understand transnational call centres (Boussebaa, Sinha, & Gabriel, 2014; Pal & Buzzanell, 2013). *Post-colonialism* is an epistemology emerging from theorists in the global south (Nayar, 2010, 2015) who argue that current relations between states are highly predicated on, and often repetitive of, earlier historical periods of colonialism (meaning the conquest by Europe and other powers of many parts of Latin America, Africa, and Asia). The contemporary period (occurring after states have achieved their independence, hence *post*-colonial) is characterized by more indirect, and some argue more insidious, power relations. This includes the continued deprivation of basic necessities among peoples in the global south, and the alignment of their elite institutions with those of the United States and other global north countries. In light of this, what is so remarkable about the global call centre industry is how it reveals the enactment of post-colonialism – directly and in action.

The post-colonialism of call centres is reflected in their ongoing dynamics and day-to-day operations. Case studies have uncovered virulent rhetoric by some global north consumers, who reify the status of workers in global south countries in post-colonial terms (*uncivilized*, *ignorant*, etc.). State leaders are known to repeat this imagery in their political discourse; to ignite nationalism in their local campaigns, they vilify outsourcing and its associated workforces. A particularly resonant marker of post-colonialism is the language of "terrorism" and "terrorists" (Nayar, 2015). This arose after the attacks of September 11, 2001 as a commonplace slander on the telephone by customers in the United States and the United Kingdom against workers in India and Muslim-majority Pakistan (Zaidi & Poster, forthcoming). In this volume, accordingly, we explore the impact of these post-colonial regroupings on the dynamics of call centres in varying contexts around the world.

Integrating Labour and Nation

Labour and nation are rarely viewed interactively. In much of the literature the exercise of nationalism is conceptualized separately from participation in the labour market. Nationalism has been the scholarly subject of political activities, the state, civil society, and even the home. Labour has been viewed as primarily a domain of spheres such as the economy, capitalism or socialism, market forces, organizational structures, and unions. The connections between labour and nation, however, are increasingly relevant in contemporary contexts.

For instance, we see the rise of specialized industries and their workforces that serve to protect nations (such as armies and associated war industries) and police the borders of nations. The world's largest single employers are military organizations (even ahead of Walmart); number one is the U.S. Department of Defense (at 3.2 million workers), and number two is the People's Liberation Army of China (at 2.3 million) (McCarthy, 2015). The first two decades of the 2000s have witnessed a massive proliferation of border construction, reinforcement, and maintenance. In turn, legions of workers are being recruited for migration management, detention, and deportment (Mezzadra & Neilson, 2013).

More recently, scholars have noted the growing securitization of many spheres of social life within nations. Poster's analysis of the nascent field of cyber-security charts many new jobs involved in creating and enforcing surveillance infrastructures, from cyber-czars, to designers of online war games, and airport screeners (2012; 2016). Cyber-spies, like call centre workers, engage in national identity management, posing from the United States as militants in countries like Iraq and Afghanistan in order to detect potential threats.

These direct employees for the military should not overshadow what may be the even greater numbers of indirect employees who support the military-industrial complex. As Enloe (2014) documents in her classic book *Bananas, Beaches, and Bases*, such labour is often highly gendered, including many types of work done by women – prostitution on military bases, nursing and health care of soldiers, and the unpaid tasks of diplomatic wives. On a broader scale, many seemingly mundane state institutions participate in the creation of ideal citizens. Employees involved in social assistance, youth training programs, or services for immigrant integration all do the daily work of making national publics.

The connections between labour and nation are similarly manifest in the transnational industries that move things across borders: people, firms, services. Nationhood becomes critical for labour in many transnational contexts, such as (1) immigration, where migrant domestic and health-care workers (Ehrenreich & Hochschild, 2003) encounter racial discrimination from employers and publics abroad, while being received as national heroes at home; (2) multinational firms and export production, where global labour confronts and threatens the localized ideals of nationalized femininity (Lynch, 2007; Ong, 1987); and (3) global labour unionism, where national contexts may challenge but also make possible the forging of international labour movements (Evans, 2014).

The transnational service sector is an emblematic site to see the interactions of nation with race, ethnicity, and class. For instance in a study of the hotel sector in the United Kingdom, Dyer, McDowell, and Batnitzky (2010) note that the assignment of cleaning and front-of-house jobs is closely tied to workers' gender, ethnicity, and migration histories. They find that "attributes based on supposed national characteristics construct workers as more or less eligible for different types of work" (p. 653). White migrants are perceived as "cosmopolitan" and given the front-office jobs in hotels, while migrant workers of colour (often from Africa and the Caribbean) are seen as most suitable for doing repetitive and unpleasant cleaning work. Workers from certain countries are constructed as hard-working and passive, while others are perceived as having propensities to make demands and therefore less suitable. Both customers and employers play a role in this process: customers evaluate the quality of service in the light of pre-existing stereotypes that are racialized and sexualized, and employers hire workers who will be successful in satisfying customer needs.

These studies highlight the embodied nature of service work and demonstrate the inextricable link between the provision of service and the enactment of identities, including national identities. Yet, in such cases the transnationalization of service work has been most frequently conceptualized in relation to the physical migration of labour or customers across national boundaries. Workers, many of whom are women, immigrate as permanent or temporary migrants to fill largely precarious jobs in domestic, care, nursing, cleaning, or retail sectors of wealthy nations. Travellers cross national borders and engage in tourism, shopping, and sex.

While call centre work shares many features with these other interactive service jobs, it is unique in an important way: the grounded location of both workers and customers in their "home" countries. Thus, rather than the workers migrating, it is the work that migrates. This represents what Aneesh (2006) refers to as "virtual migration." In this dynamic, workers travel to the nations of consumers through the telephone and the Internet, in other words virtually. The two parties do not meet in person, face to face, but rather voice to voice, as mediated by information communication technologies. Global industries, in turn, maintain the same inclusion of cheap labour forces, but without having to endure the messiness of employing migrant workers within their borders.

Recasting the Citizen Worker

If call centre employees are workers for the nation, they do so in a new capacity that straddles the local and the global. This requires recasting the historic concept of the *citizen worker*. We use this framing to propose a different vision of the interplay of nation and labour in the era of transnational customer services.

A dominant paradigm in twentieth-century United States and Europe (in theory, if not in practice) was that all workers should be citizens, and all citizens should be workers (Mezzadra & Neilson, 2013). Large sectors of the labour force worked directly for the state. National governments in many countries have been administrators of labour unions. Today some of these general principles of the citizen worker still apply: on the one hand, labour is the medium through which immigrants can achieve citizenship (for example, work visa programs that enable eligibility for residency "green cards"); and, on the other hand, disbursement of state benefits to citizens can be dependent upon particular types of labour (that is, "workfare" programs). On the whole, however, this model is clearly breaking down. The social contract between the two actors is being dissolved given that the labour market is not dependent upon "legal" workers, and alternatively not all citizens are granted rights to employment.

Our argument is that this citizen worker is being reconstituted within the transnational service industry. Workers are now labouring for the state virtually rather than in person or even from within its borders. In the process they are not only *working for* the nation, they are actively *constructing* the nation itself. This labour is different from manual, bureaucratic, or even knowledge work; it is ambassadorial and entrepreneurial. His or her role is in acting as a state representative who is capable of eliciting contracts and other tasks. It is also imaginative. Employees are creating the nation symbolically and idealistically for consumers on the telephone, through a digital world that operates in inter-country space on the channels of information and communication technology (ICT).

Our concluding chapter will describe how the citizen worker is now transforming into a *global citizen worker*. As a service employee who has direct contact with the outside world, she or he is expected to present a cosmopolitan savvy to foreign consumers and corporate clients. We will also note the drawbacks that go hand in hand with this role. While some workers experience benefits in status from being valued as global citizens, others may experience a distantiation from state entitlements as a repercussion of neoliberal policies. Still others experience a number of stigmatizations in the call centre, related to territoriality, criminality, and deportation, as Luis Pedro Meoño Artiga will show in chapter 6.

Border Work

At the heart of nation building in call centres (as elsewhere) is the creation of state borders. Conventional definitions of the *border* suggest a line drawn on a map to demarcate discrete sovereign territories or walls between nations. Scholars note that borders increasingly have functions that are flexible and dynamic (Mezzadra & Neilson, 2013). They have dual tendencies of dividing and connecting states; capacities of hierarchizing and stratifying; and "devices of inclusion that select and filter people" as well as techniques of exclusion (p. 7). In turn, nations are concerned with these sites for border reinforcement and border crossing.

As enactors of banal nationalism, workers in global call centres participate in these border activities. However, this is not the border work described above in terms of physical surveillance and obstructions of mobile bodies. Instead, the border work of call centres occurs for and through the communicative and service economies. It is a continually shifting act to promote the interests of many sets of actors surrounding global call centres, including their overseas clients and customers, national officials, and on-site managers.

This means that it has some unusual patterns, contexts, sources, and dynamics. The border work of global call centres is done as follows:

(a) *Linguistically.* Rather than being enforced through material items like paperwork, bureaucracies, checkpoints, militaries, and gunpoints, the border is constructed mentally. It takes the form of intangible sounds, words, and intonations. It is done over the telephone, via ICTs, and through the interaction itself.

(b) *By everyday people.* Rather than being carried out by official representatives of the state, it is done by workers and consumers. These are regular people who are often thought of as the users, end points, or "objects" of border control. In quite a reversal, the work of the virtual migrant is not to pass through the border but to enact, develop, imagine, articulate, sustain, and then recreate it. Their jobs have little to do directly with state or nation (although national policies on foreign direct investments, labour laws, and/or migration may deeply affect their work). Unlike the examples above, these are not workers in industries related to borders, militaries, armaments, security, etc.; they are white-collar office workers in industries of consumer sales, credit, telecommunications, etc.

(c) *By actors in other countries* (often not the state of which they are a
citizen). Mezzadra and Neilson (2013) note that "border struggles …
are played out in many contexts, often far away from geographic
borderlands" (p. 280). In the case of call centres this is in fact done
by global south actors in former colonies, who are making the
nation for corporate actors in imperial centres or metropoles.

Yet, the link between transnational capital and border work has been
longstanding. One might see the border work of call centres as a twenty-
first century version of the twentieth-century maquiladora. Maquiladoras are multinational factories that operate on the border between the
United States and Mexico. Made possible by free trade agreements of
the 1970s and 1980s, they literally have straddled the border in some
cases: conveyer belts move parts manufactured by cheap workers in
Mexico to higher-costing assemblers in the United States. In the current
era, nations no longer need such elaborate architectures or the side-by-
side positioning of other nations for this global coordination to operate. Workers and customers across faraway nations are in contact with
one another virtually. As a result the boundaries between nations are
enacted on a continuous basis as customers and workers cross national
borders with every call.

Throughout this volume we will document what Mezzadra and Niel-
son describe as "border as method" – that is, the border as a method
for capital, selectively opening and closing, filtering and excluding
different kinds of workers, and stretching wide or narrowing around
economic corridors and zones. In our case the border is a method for
everyday citizen workers in call centres. These are employees who (in
the interests of call centre stakeholders) redefine the meaning of the
nation on a regular basis, highlight some features of the nation over
others, recast their own nationality, move the lines on the map to appeal
to consumers, and cross the border(s) daily in their virtual labour.

Multiplicities and Hybridities of Nationhood

The border work of call centres complicates our understanding of
nation even further. As a crossroads between many countries through
virtual infrastructures, the labour of nation making involves a number
of multiplicities and hybridities.

To start with, the labour of the new citizen worker is multiplicative.
Workers are creating not only one nation but *many nations at the same time.*

For the client firm and its consumers, they use linguistic skills to create nations in the global north. For the outsourced firm and local state offi-cials, they create the nation where the call centre is located in the global south. Being a labourer on the virtual border is to do the work of both sides simultaneously.

Thus, in call centres borders are performed and contested in the light of multiple nationalistic pressures. This is compounded by the fact that call centres are increasingly taking contracts from more than one coun-try at a time, whether in the same core language or in many linguistic groupings. In turn, workers are often asked to inhabit and display these multiple, discrete nationalities and then enact corresponding borders. This is an application of what Mezzadra and Neilson (2013) refer to as the "heterogeneity" of borders and their labours in the contemporary era.

In addition, call centres are very representative of hybrid identities. As Bhabha (1994) describes, hybridity is a common experience of peo-ples in post-colonial societies as they negotiate identities from both their colonizers and anti-colonial movements. Furthermore, many the-orists have noted that the nation itself is a hybrid construct, constantly emerging and re-emerging from many sources at once.

Along these lines there are many hybridities of call centre labour. The work often requires dualized skills and involves dualized tasks. Specific sets of linguistic resources are needed for the labour of alter-nating between nations. Cases from El Salvador and the U.S.-Mexican border in this volume illustrate the common use of code switching, in terms of flipping back and forth from Spanish to English. Further-more, the inherent multiplicity of global call centres leads employers to search for hybrid people. In their recruiting, many call centres tar-get specialized communities of personnel who have pre-existing dual identity resources of the colony and the metropole. In fact, this is how many U.S. researchers have gained access to Indian call centres as a site of study; employers are eager to hire foreign nationals residing in the United States who are familiar with India but can teach accent and culture to local populations in India. (However, this can backfire: see Aneesh, 2015, for a humorous account of an Indian-born researcher liv-ing in the United States who was told by the call centre that his accent was "too American.") Some global call centres (in countries like Gua-temala and El Salvador) make it their *primary* aim to hire workers with lived experience in the countries they are servicing. Consequently their workforces are comprised largely of migrants. In this sense, recon-necting diasporic and former nationals to their homeland is a tactical recruitment strategy. These employers then benefit from the ongoing personal ties that workers have to their families abroad, and the continual investments and enhancements of their social capital in hybrid national identities.

Fundamentally, the hybrid space of the call centre can have detrimental effects on workers. To be on the borderland, as Anzaldúa (1999) illuminates so brilliantly, is to be nowhere and everywhere. For the Chicanas and Chicanos in her example, the borderland is neither completely Mexico nor the United States. Its inhabitants learn to become a part of both worlds while not being accepted fully by either. In this volume we examine the fallout (and tensions) of border making and crossing on everyday people. Many workers (from countries like El Salvador, Guatemala, and Morocco, as this volume will show) are migrants and former victims of physical borders, yet they are asked to enforce those same state borders within cross-national telephone conversations. Call centres demonstrate the contradictions of virtual space within the transnational service industry.

Aims, Questions, Themes

The overall aim of this collection is to provide an interdisciplinary perspective on the issue of nationhood and transnational customer service from the fields of sociology, education, labour, geography, anthropology, linguistics, humanities, business, management, and organizational behaviour. We embrace multiple methodologies: qualitative and quantitative, interpretive and content analysis, surveys, case studies, and ethnography. The book brings together leading scholars who are doing ethnographic work on offshored transnational service work in many parts of the world but especially in the global south, particularly Mauritius, Morocco, the Philippines, Guatemala, El Salvador, Guyana, and the U.S.-Mexico border.

Borders in Service explores the many ways that workers engaged in cross-national service exchanges create imaginaries of nations, both their own and those in which their customers are based. The following overarching questions are addressed: How are the ideas of nation, nationhood, and nationalism manifested in cross-national service exchanges? How are nationhood and the national constituted? What engages and activates the nation as a narrative and a practice? When is nationalism emphasized, and when is it suppressed? How are citizens and outsiders constructed during service encounters? How do histories and geographies have an impact on the enactment of various cross-national service interactions? How do the relationships between gender, race, class, sexuality, and nation affect service encounters? Critically, how does the national intersect with the transnational in rhetoric, imageries, and practices? And how are the boundaries of national belonging, citizenship, and value formed, enacted, and contested?

The chapters examine a wide range of dynamics within global call centres and how they reflect national imperatives. They explore nationhood and national identity in the various material and symbolic elements of these settings. Contributors, for instance, study the interactional and linguistic features of call centres: language, speech, and accent issues; conversational dynamics between workers and consumers; and expressions of consumer sentiment. They look at the informal aspects of work culture and space, like the physical space, including arrangements, architecture, decor, and the corporality of workers, including their bodies, comportment, and dress. They examine the formal aspects of the labour process: training processes and content; organizational, human resource, and labour policies; workplace stratification systems; and unionization and resistance. Finally, they consider the socio-economic aspects of the broader context and communities outside the call centre: the intersections of labour with gender, race, ethnicity, class, and sexuality; the representations of call centres in the media (news, television, film, Internet); and discussions of call centres within state policies, religious bodies, educational institutions, community associations, and civil society.

We argue that the significance of this exercise is initially one of scope. Our fundamental goal is to show the geographic dispersion of the call centre industry beyond India. We undertake one of the first comparative examinations of call centres in the dawn of this transnational industry that pays special homage to the meaning and significance of national difference. In an interdisciplinary field that tends to emphasize similarities we illuminate the profound variations of call centre labour patterns arising from the socio-political context. The authors in this volume provide rich ethnographic detail of particular geographies, affirming that Central American, South American, African, and Asian landscapes make highly divergent experiences for call centres. Critically, when most call centre research collapses regions of the world, or remains focused on the global north, our authors emphasize the urgency for research on the global south, as well as specific points within it. We also show how the industry itself is re-centring, further attenuating from the global north and shifting towards the global south. It is here that the transnational call centre industry is increasingly being networked and organized, and that authority relations are re-situating.

Theoretically, this volume challenges a number of premises and unquestioned assumptions of dynamics in the call centre industry. We seek to dispel a variety of myths about the outsourcing of call centres (as well as many other types of transnational businesses): (1) that multinational firms are placeless and/or will dissolve national boundaries; (2) that technology will override discrepancies of place and geography; (3) that the skills used in call centres are universally transferable, as theories such as human capital would predict; and (4) that call centre workforces are more or less monolithic without any significant internal differentiations or stratifications, local or otherwise.

In contrast, we present many counter-trends and alternative readings of the way the global call centre industry is advancing, by placing nation and nationhood at its core. Throughout the chapters this volume emphasizes four themes: (1) that *national identity management* and authenticity work are spreading around the globe, rather than receding or remaining specific to the case of India; (2) that nationhood is often inextricably tied to the way call centres define themselves through a process of *national branding*, which is central to their marketing and promotion strategies for attracting foreign clients and their migrant workforces abroad (that is, as a homecoming); (3) that the *global citizen* is emerging as the model worker for international-facing industries, as one who interacts with foreign personnel and presents a cosmopolitan image while still embodying national symbols and practices in other contexts; and (4) that *communication hierarchies* are often the underlying foundation for global call centre labour processes, as they differentiate the workforce along transnational lines and linguistic capacities.

Outline of the Book

The chapters are organized thematically in three sections. Before considering how nationalism affects the operations of call centres and their workers, we reflect on how call centres play a role in creating the image of the nation. Part I looks at the national brands that are generated by state governments in El Salvador, Guyana, and Mauritius through policies concerning the workforces of transnational call centres.

In chapter 2, Cecilia Rivas explores the construction of the brand "El Salvador Works," which situates call centre employees as hyper-mobile, global workers whose skills are ripe for export and circulation within the global economy. Branding by the state investment agency involves the naming of Salvadorans as hard workers who are fluent in English as well as "neutral Spanish." Yet, as her ethnography reveals, "the ideal, bilingual Salvadoran call centre agent does not exist naturally as an immediately available workforce." Rather, English fluency is associated with diasporic Salvadorans who are "returnees," either because they have been deported from the United States as a result of criminal convictions or because they are of Salvadoran heritage. In an attempt to dislodge the association of ideal call centre workers with the supposed criminality of bilingual deportees, the "El Salvador Works" brand instead utilizes a "Meet Your Roots" campaign in which diasporic Salvadorans are encouraged to visit and discover their "roots" while working in a call centre. In this way the national branding exercise involves policies, organizational practices, and programs through which ideal call centre workers are actually created. Nation branding involves the construction of not only labour but also geographies.

As chapter 3 on Guyana shows, the call centre sector facilitates nation branding not only in terms of the promotion of the nation as an ideal location for foreign direct investment but also in terms of the marketing of call centre jobs as ideal for local, middle-class, cosmopolitan citizens. As Trotz, Mirchandani, and Khan's call centre interviews reveal, employees' experiences of their work (which include low pay, routinized jobs, harsh penalties, and little protection from labour law violations) disrupt these state discourses. Neoliberal state policies and the proliferation of labour market precarity go hand in hand. The chapter explores three forms of sovereignty that are continually being enacted and expressed within Guyanese call centres – state sovereignty, customer sovereignty, and the sovereignty of global capital. Within nation-branding discourses, cheap labour is constructed as a national advantage through which foreign investors can be encouraged to set up call centres in Guyana. Yet multinational investors demand educated, cosmopolitan workers, who are forced to become the promised cheap labour, because of limited employment alternatives.

In similar ways, as Benner and Rossi show in chapter 4, the branding of Mauritius involves national constructions that convey "multiple incarnations of itself to assert connections to, and facilitate relations with, other places within a competitive geography." Although physically located close to South Africa and India, and a member of the African Union, the country is branded as an ideal location for call centre work given its diverse population that is deeply embedded within diasporic networks, and its location off the coast of Australia. The chapters in this section reveal that branding plays a key role in the context of global competition for call centre contracts. The call centre sector is deeply shaped by the exercise of state-oriented nationhood.

After considering the way in which national call centre workforces are advertised to the world, part II explores more closely the development of nationally appropriate workers. It considers how nation states and call centre industry leaders are heavily invested in the social reputations of these workforces. In many instances they dictate the parameters of how those employees should behave, or how the public should view them. Some of these imageries are uplifting, like the elevation of the call centre worker as a "national hero" in the Philippines. Others are deflating, like the shameful "gang members" in Guatemala. The experience of such contradictory narratives for employees is considered in this section.

In chapter 5, Salonga analyses two commercials by Convergys, a prominent transnational call centre in the Philippines, and traces how call centre workers are constructed as national saviours, helping both their families and the nation through their engagement in demeaning and arduous work. Such hero nationalism is based, however, on racist and culturalist constructions of Filipinos as "naturally caring and nurturing" and having a "sweet, caring" communicative tone that facilitates their success as call centre workers. Chapter 6 explores the case of Guatemala. Given that young, middle-class, bilingual (English and Spanish) students populate call centres, the presence of U.S. deportees (called "homies") who possess English-speaking skills that are perfectly suited to the call centre industry results in a situation of "exclusionary nationalism." Such nationalism deepens ethnic identities, rather than unifying citizens under the banner of a common nation. The call centre sector becomes a site for creating a "nation within the nation"; while working alongside one another, employees of one group are constructed as outsiders (potential gang members and drug addicts), while the others are representatives of the nation. As Meoño Artiga argues, call centres are "real and imaginary urban spaces" within which workers compete to be the "new Guatemalans – flexible, cosmopolitan and consumerist."

Part III reflects on the production of nation within the labour process itself by workers for their employers. Here we see how workers engage in performances of nation and border crossing while they are on the telephone with their customers. In the process, we see that language is one of the key markers of nationalism within global call centres; wording, accent, and voice become pivotal dimensions for defining the nation when a firm operates across state lines. These chapters delve into the complex processes by which multiple languages are deployed in call centres, particularly when those organizations lie on or close to political boundaries (like the United States–Mexico case in chapter 8) and when they serve many linguistic consumer populations at once (like the Moroccan case in chapter 7).

In chapter 7, Elmoudden traces the ways in which call centre workers discursively cross national spaces for their work, and one worker in Morocco notes, "Every time I leave the centre, I feel as if someone is taking my passport away from me and I will have to re-enter Morocco." Such honorary immigrants hold neither immigration rights nor the security of citizenship. Workers engage in discursive crossings through language in the context of multilingual transnational call centres, where agents speak mostly French but also Spanish, English, and German in order to serve clients in Europe. In chapter 8 the discussion focuses on the discontinuities between the construction and experiences of work that necessitate the intricacies and tensions of performing hybrid national identities – that is, one for the local community and another for the customers on the telephone. Heyman and Alarcón note that nationality in global call centres is enacted as a "border performance." In other words, workers are required to engage in deliberate routines to display and exhibit nationalism properly, capitalizing on their "Spanish heritage," which forms the basis of their appropriateness for jobs. Serving Spanish- and English-speaking clients, the bilingualism is seen to arise out of the workers' location in El Paso, rather than being conceptualized as an organizationally valuable skill.

The conclusion in chapter 9 reviews the overall argument on the expression of banal nationalism in global call centres, and how it relates to structural patterns of geographic decentering in the global service economy. It delineates the four features of nation and nationhood that appear throughout the volume: expansions of national identity management; the increasing role of nation branding; the making of global citizen workers; and the resilience of communicative hierarchies along national markers. It also reviews the tensions that these dynamics create for workers (and employers), and the kinds of resistance that appear.

Collectively the essays show that nationhood is exercised within the call centre sector much more deeply, and in much more varied ways, than previous studies have considered (especially regarding the single focus on India). Despite international standards, standardized labour processes, and common technologies within transnational call centres around the world, the actual experience and construction of work differ significantly. Nation, national identity, and nationalism are critical elements of the social construction of global call centres. We will show that borders in service are increasingly relevant as states assert their agendas through the call centre industry and, alternatively, as executives benefit from instilling a sense of nationhood and nationality in their workers.

REFERENCES

Anderson, B. (1991). *Imagined communities*. New York, NY: Verso.
Aneesh, A. (2006). *Virtual migration*. Durham, NC: Duke University Press. http://dx.doi.org/10.1215/9780822387534
Aneesh, A. (2015). *Neutral accent: How language, labor, and life become global*. Durham, NC: Duke University Press.
Anzaldúa, G. (1999). *Borderlands – La Frontera: The new Mestiza*. 2nd ed. San Francisco, CA: Aunt Lute Books.
Arun, M.G. (2013, 25 March). Last call: India's about to hang up on call centre culture. *India Today*. Retrieved from http://indiatoday.intoday.in/story/bpo-industry-call-centre-culture-dying-in-india/1/258032.html
Basi, T.J.K. (2009). *Women, identity and India's call centre industry*. London, UK: Routledge.
Batt, R., Holman, D., & Holtgrewe, U. (2009). The globalization of service work: Comparative institutional perspectives on call centers. *Industrial & Labor Relations Review, 62*(4), 453–88. http://dx.doi.org/10.1177/001979390906200401
Benner, C. (2006). "South Africa on-call": Information technology and labour market restructuring in South African call centres. *Regional Studies, 40*(9), 1025–40. http://dx.doi.org/10.1080/00343400600928293
Bhabha, H.K. (1994). *The location of culture*. London, UK: Routledge.
Billig, M. (1995). *Banal nationalism*. London, UK: Sage Publications.
Boussebaa, M., Sinha, S., & Gabriel, Y. (2014). Englishization in offshore call centers: A postcolonial perspective. *Journal of International Business Studies, 45*(9), 1152–69. http://dx.doi.org/10.1057/jibs.2014.25
Brophy, E. (2010). The subterranean stream: Communicative capitalism and call centre labour. *Ephemera, 10*(3/4), 470–83.
Burawoy, M., Blum, J.A., George, S., Gille, Z., & Thayer, M. (2000). *Global ethnography: Forces, connections, and imaginations in a postmodern world*. Berkeley, CA: University of California Press.

Burgess, J., & Connell, J., eds. (2006). *Developments in the call centre industry: Analysis, changes and challenges.* London, UK: Routledge.

Das, D., Dharwadkar, R., & Brandes, P. (2008). The importance of being "Indian": Identity centrality and work outcomes in an off-shored call center in India. *Human Relations, 61*(11), 1499–530. http://dx.doi.org/10.1177/0018726708096636

D'Cruz, P., & Noronha, E. (2014). The interface between technology and customer cyberbullying: evidence from India. *Information and Organization, 24*(3), 176–93. http://dx.doi.org/10.1016/j.infoandorg.2014.06.001

Dean, J. (2009). *Democracy and other neoliberal fantasies: Communicative capitalism and left politics.* Durham, NC: Duke University Press. http://dx.doi.org/10.1215/9780822390923

Deery, S., & Kinnie, N. (Eds.). (2004). Introduction: The nature and management of call centre work. In *Call centres and human resource management : A cross-national perspective* (1–22). New York, NY: Palgrave Macmillan.

Dyer, S., McDowell, L., & Batnitzky, A. (2010). The impact of migration on the gendering of service work: The case of a west London hotel. *Gender, Work and Organization, 17*(6), 635–57. http://dx.doi.org/10.1111/j.1468-0432.2009.00480.x

Ehrenreich, B., & Hochschild, A.R. (Eds.). (2003). *Global woman: Nannies, maids, and sex workers in the new economy.* New York, NY: Metropolitan Books.

Elden, S. (2013). *The birth of territory.* Chicago, IL: University of Chicago Press. http://dx.doi.org/10.7208/chicago/9780226041285.001.0001

Enloe, C. (2014). *Bananas, beaches and bases: Making feminist sense of international politics* (2nd ed.). Berkeley, CA: University of California Press.

Evans, P. (2014). *National labor movements and transnational connections: Global labor's evolving architecture under neoliberalism.* UC Berkeley Institute for Research on Labor and Employment, Working Paper no. 116-14. http://irle.berkeley.edu/workingpapers/116-14.pdf.

Fox, J.E., & Miller-Idriss, C. (2008). The "here and now" of everyday nationhood. *Ethnicities, 8*(4), 573–6. http://dx.doi.org/10.1177/14687968080080040103

Free, A. (2014, June). "Development," profiles and prospects: Labour in Kenya's outsourced call centres. *Critical African Studies, 1392,* 1–19.

Friedman, T.L. (2005). *The world is flat: A brief history of the twenty-first century.* New York, NY: Farrar, Straus, & Giroux.

Glucksmann, M. (2004). Call configurations: Varieties of call centre and divisions of labour. *Work, Employment and Society, 18*(4), 795–811.

Graham, M., & Mann, L. (2013). Imagining a silicon savannah? Technological and conceptual connectivity in Kenya's BPO and software development sectors. *Electronic Journal of Information Systems in Developing Countries, 56*(2), 1–19.

Grandey, A.A., Fisk, G.M., & Steiner, D.D. (2005). Must "service with a smile" be stressful? The moderating role of personal control for American and French employees. *Journal of Applied Psychology, 90*(5), 893–904. http://dx.doi.org/10.1037/0021-9010.90.5.893

Grandey, A.A., Rafaeli, P.A., Ravid, M.S., & Wirtz, P.J. (2010). Emotion display rules at work in the global service economy. *Journal of Service Management, 2*(3), 3.

Holman, D., Batt, R., & Holtgrewe, U. (2007). *The global call center report.* Ithaca, NY: ILR School, Cornell University.

Hunter, M., & Hachimi, A. (2012). Talking class, talking race: Language, class, and race in the call center industry in South Africa. *Social & Cultural Geography, 13*(6), 551–66. http://dx.doi.org/10.1080/14649365.2012.704642

Huws, U. (2009). Working at the interface: Call-centre labour in a global economy. *Work Organisation, Labour & Globalisation, 3*(1), 1–8. http://dx.doi.org/10.13169/workorgalaboglob.3.1.0001

Lacey, M. (2005). Accents of Africa: A new outsourcing frontier. *New York Times.* http://www.nytimes.com/2005/02/02/business/worldbusiness/02outsource.html

Lan, P.-C. (2001). The body as a contested terrain for labor control: Cosmetics retailers in department stores and direct selling. In R. Baldoz, C. Koeber, & P. Kraft (Eds.), *The critical study of work* (83–105). Philadelphia, PA: Temple University Press.

Lee, D. (2015, 1 February). The Philippines has become the call-center capital of the world. *Los Angeles Times.* Retrieved from http://www.latimes.com/business/la-fi-philippines-economy-20150202-story.html

Lynch, C. (2007). *Juki girls, good girls.* Ithaca, NY: ILR Press.

Mann, L., Graham, M., & Friederici, N. (2014). *The Internet and business process outsourcing in East Africa.* Oxford, UK: Oxford Internet Institute.

McCarthy, N. (2015, June 23). The world's biggest employers. *Forbes.* http://www.forbes.com/sites/niallmccarthy/2015/06/23/the-worlds-biggest-employers-infographic/#3f338e2a51d0

Mezzadra, S., and Neilson, B. (2013). *Border as method, or the multiplication of labor.* Durham, NC: Duke University Press. http://dx.doi.org/10.1215/9780822377542

Mirchandani, K. (2012). *Phone clones.* Ithaca, NY: Cornell University Press.

Mutz, D.C., & E. Mansfield. (2011). Us vs. them: Mass attitudes towards offshoring and outsourcing. Paper presented at annual meeting of American Political Science Association, Seattle, WA, 1–4 September.

Nadeem, S. (2011). *Dead ringers.* Princeton, NY: Princeton University Press. http://dx.doi.org/10.1515/9781400836697

Nayar, P.K. (2010). *Postcolonialism: A guide for the perplexed*. London, UK: Continuum International.

Nayar, P.K. (Ed.). (2015). *The postcolonial studies dictionary*. Chichester, UK: Wiley Blackwell.

Noronha, E., & D'Cruz, P. (2009). *Employee identity in Indian call centres*. Thousand Oaks, CA: Sage.

Ong, A. (1987). *Spirits of resistance and capitalist discipline: Factory women in Malaysia*. Albany, NY: State University of New York Press.

Pal, M., & Buzzanell, P.M. (2013). Breaking the myth of Indian call centers: A postcolonial analysis of resistance. *Communication Monographs, 80*(2), 199–219. http://dx.doi.org/10.1080/03637751.2013.776172

Patel, R. (2010). *Working the night shift*. Palo Alto, CA: Stanford University.

Poster, W.R. (2002). Racialism, sexuality, and masculinity: Gendering "global ethnography" of the workplace. *Social Politics, 9*(1), 126–58. http://dx.doi.org/10.1093/sp/9.1.126

Poster, W.R. (2007). Who's on the line? Indian call center agents pose as Americans for U.S.-outsourced firms. *Industrial Relations, 46*(2), 271–304.

Poster, W.R. (2012). The case of the U.S. mother / cyberspy / undercover Iraqi militant, or how global women have been incorporated in the technological war on terror. In R. Pande, T. van der Weide, & N. Flipsen (Eds.), *Globalization, technology diffusion, and gender disparity: Social impacts of ICTs* (247–60). London, UK: Routledge. http://dx.doi.org/10.4018/978-1-4666-0020-1.ch020

Poster, W.R. (2015). Sound bites, sentiments, and accents: Digitizing communicative labor in the era of global outsourcing. Presentation at the Annual Meeting of the Society for the Social Studies of Science. Denver, CO.

Poster, W.R. (2016). Women body screeners and the securitization of space in Indian cities. In S. Raju & S. Jatrana (Eds.), *Women workers in metro cities of India* (312–32). New Delhi, India: Cambridge University Press.

Poster, W.R., & Yolmo, N.L. (2016). Globalization and outsourcing. In S. Edgell, H. Gottfried, & E. Granter (Eds.), *The Sage handbook of the sociology of work and employment* (576–96). London, UK: Sage Publications.

Qiu, J.L. (2010). Network labour and non-elite knowledge workers in China. *Work Organisation, Labour & Globalisation, 4*(2), 80–95. http://dx.doi.org/10.13169/workorgalaboglob.4.2.0080

Rowe, A.C., Malhotra, S. & Pérez, K. (2013). *Answer the call: Virtual migration in Indian call centers*. Minneapolis, MN: University of Minnesota Press. http://dx.doi.org/10.5749/minnesota/9780816689385.001.0001

Russell, B. (2008). Call centers: A decade of research. *International Journal of Management Reviews, 10*(3), 195–219. http://dx.doi.org/10.1111/j.1468-2370.2008.00241.x

Russell, B. (2009). *Smiling down the line: Info-service work in the global economy*. Toronto, ON: University of Toronto Press.

Sassen, S. (2006). *Territory, authority, rights*. Princeton, NJ: Princeton University Press.

Skey, M. (2011). *National belonging and everyday life: The significance of nationhood in an uncertain world*. Basingstoke, UK: Palgrave Macmillan. http://dx.doi.org/10.1057/9780230353893

Sonntag, S.K. (2005). Appropriating identity or cultivating capital? Global English in offshoring service industries. *Anthropology of Work Review*, 26(1), 13–20. http://dx.doi.org/10.1525/awr.2005.26.1.13

Taylor, P., & Bain, P. (2005). "India calling to the far away towns": The call centre labour process and globalization. *Work, Employment and Society*, 19(2), 261–82. http://dx.doi.org/10.1177/0950017005053170

Thelen, S.T., Honeycutt, E.D., Jr., & Murphy, T.P. (2010). Services offshoring: Does Perceived service quality affect country-of-service-origin preference? *Managing Service Quality*, 20(3), 196–212. http://dx.doi.org/10.1108/09604521011041943

Thelen, S.T., Thelen, T.K., Magnini, V.P., & Honeycutt, E.D., Jr. (2009). An Introduction to the offshore service ethnocentrism construct. *Services Marketing Quarterly*, 30(1), 1–17. http://dx.doi.org/10.1080/15332960802467565

Thelen, S.T., Yoo, B., & Magnini, V.P. (2010). An examination of consumer sentiment toward offshored services. *Journal of the Academy of Marketing Science*, 39(2), 270–89. http://dx.doi.org/10.1007/s11747-010-0192-7

Thite, M., & Russell, B. (Eds.). (2009). *The next available operator: Managing human resources in Indian business process outsourcing industry*. New Delhi, India: Sage. http://dx.doi.org/10.4135/9788132101260

Wacquant, L. (1995). Pugs at work: Bodily capital and bodily labour among professional boxers. *Body & Society*, 1(1), 65–93. http://dx.doi.org/10.1177/1357034X95001001005

Warhurst, C., & Nickson, D. (2007). Employee experience of aesthetic labour in retail and hospitality. *Work, Employment and Society*, 21(1), 103–20. http://dx.doi.org/10.1177/0950017007073622

Witz, A., Warhurst, C., & Nickson, D. (2003). The labour of aesthetics and the aesthetics of organization. *Journal of Composite Materials*, 10(1), 33–54.

Zaidi, Y. & Poster, W.R. (forthcoming). Shifting masculinities in the South Asian outsourcing industry: Hyper, techno or fusion? In H. Peterson (Ed.), *Gender in Transnational Knowledge Work*. New York, NY: Springer.

PART ONE

Call Centres as Building Blocks for Narratives of the Nation State

2 "El Salvador Works": The Creation and Negotiation of a National Brand and the Transnational Imaginary

CECILIA M. RIVAS

Introduction

"Hasta caemos mal de amables" (we are polite to the point of annoying some callers), said Laura and Elizabeth to summarize their many interactions with customers at a call centre in San Salvador, the capital of El Salvador. They were sharing anecdotes about their work, and the conversation was lively. For these young women, the call centre has been a workplace in which they have achieved considerable career growth. "We feel like we have grown along with the company and in our careers," Elizabeth said. "The work environment is really good," Laura added, "and, yes, it is a young working environment, where we feel it is important to innovate, to not fall asleep!"[1]

Their remarks present a point of departure and a glimpse into the career aspirations and accomplishments of Laura and Elizabeth. Moreover, they connect the growth of call centres in San Salvador to constructions of belonging, career advancement, productivity, and innovation and to ideal qualities of Salvadoran customer service, such as politeness. The presence of this transnational worksite in the everyday life of a small yet significant group of mostly young and urban Salvadorans raises a range of questions related to social class, aspirations, and mobility and the extent to which call centres represent connectivity, connectedness, and rapid social change in post-war El Salvador.

In this chapter I take a closer look at the operations of call centres in San Salvador. Focusing on hiring processes, I analyse the construction of ideal workers and ideal citizens, along with the ironies of deported labour, language "neutrality," and deterritorialized voices. I argue that call centres are important and complex spaces in which the narratives

and ideas of Salvadoran workers and a "Salvadoran brand" are constructed and contested, nationally and transnationally. Furthermore,
the call centre is emblematic of a transnational agenda and an example
of the ongoing and deep transformations in the ways that Salvadorans relate to work, to processes of globalization, to perceptions of class
and educational level, and to new understandings of individual risk
and labour flexibility. Call centres are spaces not only of opportunity
but also of meaningful constraints and contradictions. In the context
of El Salvador's post-war social, economic, and institutional transformations, the branding and construction of the country as attractive for
foreign direct investment is intertwined with formations of language
ability (especially fluency in English) and the relative novelty of outsourced service work. Among other elements, the brand and slogan
"El Salvador Works" (created by the country's export and investment
promotion agency) is part of this story of transformative connections.
With this in mind, close attention will be paid to various facets of the
story of how call centre work is possible in El Salvador. How do Salvadorans become readily available call centre workers – and, like Laura and
Elizabeth, very polite? What are the expectations attached to this job? I
examine the ways in which the views and opinions expressed by call
centre agents and managers about this occupation intersect (and, often,
contrast) with other discourses, images, and practices of the national and
transnational.

Method

This chapter analyses interviews, advertisements, and other document
sources (such as training manuals, websites, newspaper stories, and
brochures) to contextualize and shed light on the intersection of labour
and nation in El Salvador. Examined together, these sources of information about recruitment and training practices reveal the degree to which
this intersection matters in the construction of a Salvadoran brand and
an ideal call centre labour force.

I draw from twelve in-depth qualitative interviews that I conducted
with employees and other people involved in the call centre sector; in
particular, here I focus on my conversations with the human resources
personnel of three call centres in San Salvador.[2] Personal contacts were
important in finding people to interview for this project. Laura and Elizabeth, introduced earlier, were my first informants. I had known them
for several years in El Salvador, and one of them was the first to tell me

about her intention to apply to work at a call centre that had started its operations in the country in the late 1990s. Laura and Elizabeth had been working at this call centre for approximately four years when I first approached them in early 2004, asking for their thoughts about two bilingual call centres that, according to news reports,[3] were about to be established in San Salvador. They replied that Gabriela (another young woman whom we knew) was going to work in the human resources office of one of these new call centres. I had the opportunity to conduct extensive interviews with all three of them in 2005 and to follow up with Gabriela in 2006 after her first year in the new large bilingual call centre. They were important contacts when I began my research and were instrumental in referring me to some of their co-workers, effectively putting me in touch with many potential interviewees as the research project progressed. In addition to these personal contacts and their co-workers – a snowball sample – I also contacted potential informants whose names (along with the names of their workplaces) appeared in news stories or advertisements related to the call centre sector. Since most employment advertisements include a telephone number (and sometimes an email address), it was possible to contact them for further information.

These strategies led me to meet several people who were interested in sharing their experiences of working at a call centre. Many (but not all) of them resembled Laura, Elizabeth, and Gabriela. The typical person was a woman in her twenties, bilingual, single or recently married, and with a good level of education (she had completed high school and in most cases was studying towards her university degree). In some ways, personal contacts and advertisements gave me an initial sense of the people who typically worked at the call centres. However, it was these women who also led me to understand that they were only one segment (or face) of this complex, diverse, and emerging sector. Thanks to our interviews, it became clear to me that there was no single, typical, or ideal call centre worker. Moreover, their descriptions of the workplace, and of the job applicants who aspired to belong to it, often revealed the diversity – even the contradictions – of the call centre. In the interviews, the managers and some employees conveyed various degrees of enthusiasm and satisfaction with their work environment, but they also expressed concerns or uncertainties about less attractive aspects of the job (for instance, stress and/or high turnover rates).

Along with interviews of call centre employees, I analyse business news, advertisements, and job descriptions that appeared in the employment section of major Salvadoran newspapers, especially *La Prensa Gráfica*. I also consulted relevant websites, for example that of PROESA (Agencia de Promoción de Exportaciones e Inversiones de El Salvador), the Salvadoran agency for the promotion of exports and investments. These materials signal ideas about language learning and ability, training, marketing, and foreign investment promotion; they are important as we try to understand the greater context of call centre employment, media representations of call centres in El Salvador, and the growth, promotion, and branding of this sector. Together, interviews, images, and other discourses and representations provide a snapshot of call centre work, situated in a specific time and place: San Salvador during the first decade of the twenty-first century.

"El Salvador Works"

Call centres are not unique to El Salvador, but they are connected to the imaginary of Salvadoran transnationalism in unique and significant ways. During 2004, when I began researching bilingual (English-Spanish) and monolingual Spanish call centres in San Salvador, two major companies (one North American, the other European) that had recently established their call centre operations in the country were becoming places of employment for a specific sector of Salvadorans – especially the young and fluent in English. These were neither the first nor the only call centres in San Salvador, but they were notable in that they were among the first to require English-language fluency and to recruit widely on this basis. One of the first call centres, established in 1999, was a Spanish-language call centre for a European company that had started operating in El Salvador in 1998. Several years later this call centre would continue to be presented as an important component of an ambitious investment promotion project and the marketing of the idea of Salvadoran Spanish as "neutral" in pronunciation and speech patterns. This marketing of neutral Spanish, in addition to the demands for bilingual workers, carries interesting, critical implications of linguistic ability and labour flexibility. Through an examination of the country's call centre sector, it is possible to evaluate the emergence of bilingual and transnational identities that converge with a nation-making agenda in El Salvador. This labour site is one of many versions of the imaginary of El Salvador as a nation without borders, where the national and the transnational are mutually constitutive and where distances "accentuate the needs for connection" (Rivas, 2014, p. 8). The country, in its specificity, becomes a strategic, ideal location for call centres, while it is also constructed as suitable and competitive for the everyday practices and logistics of global customer service.

Currently India and the Philippines are countries with a strong presence and global reputation in the call centre industry. Their industries not only draw on the convenience of rapid global connections but also become intensely local, highlighting certain traits or perceptions, such as hospitality, location, and familiarity with English, to build their reputations as suitable places for companies to establish their centres. For instance, the widely held notion that call centre agents in India are more proficient in the field of computers and information technology may be stereotypical, yet it feeds into customer sentiments regarding reliability and strong technical knowledge. India dominates the call centre sector globally, with two million customer-contact jobs, although there has been a shift of these jobs in recent years from India to several other countries, such as the Philippines.[4]

El Salvador's call centre sector is relatively small, not near the scale or volume of Great Britain's during the 1990s, where an estimated 247,000 to 400,000 agents were employed (Ball, 2003, p. 203). India and the Philippines are countries in which English-language education is integral to the school system; they are nations that have been intertwined with historical and colonial links to England and the United States for generations. Clearly I am drawing a contrast to countries in which this sector seems to be large scale and visible. In this chapter I am discussing a relatively new and small sector that is concentrated in a small capital city. Approximately five thousand people worked as agents in 2005–6, when I began to conduct interviews with employees at various call centres located in San Salvador. By 2012 the number of employees had nearly doubled to ninety-four hundred agents, "in forty-five local and foreign companies."[5] During 2014 some call centres were expected to grow by 30 per cent in their El Salvador-based operations; one company projected the creation of approximately eighteen hundred new agent positions.[6]

How and why are call centres present in El Salvador? Agencies such as PROESA attract foreign companies to employ Salvadoran labour and to create jobs that require fluency in English and familiarity with global consumer products and services. PROESA, among other institutions, has played a significant role in the expansion of this sector in the country. The discussion of call centres, then, involves the formation of ideas of Salvadoran labourers and their circulation of linguistic and mobility capital. The call centres in great measure exist thanks to globalizing processes that include media and communication technologies and the "ethnoscapes" (Appadurai, 1996, pp. 33–4) produced by the movements of people – in particular, the mobilities of labour and corporate needs, along with the linguistic capital of the workers. Paradoxically, these worksites are grounded in critical local conditions and opportunities such as those that investment promotion agencies (and other institutions) want to develop.

Since the end of the civil war in 1992 the Salvadoran government has pursued a strategy of privatizing public services and utilities, including ANTEL (the national telecommunications institution) and the banking and financial services sectors. These projects of privatization and investment promotion in part laid the groundwork for what is now the call centre sector. They made it legally and logistically possible for transnational telecommunications corporations (for instance, companies from France and Spain) to acquire significant ownership of the landline and wireless sectors. By 1998, foreign investors could establish their operations within newly created Salvadoran legal frameworks that regulated export-processing zones and other concerns related to outsourcing. For instance, the legislative decree 405, Ley de Zonas Francas Industriales y de Comercialización (initially passed into law by the Salvadoran legislative assembly in September 1998 and subsequently reformed), establishes the importance and responsibility of the Salvadoran government in developing a range of capacities to attract foreign direct investment. The decree outlines dispositions and guidelines to modernize and update the legal and regulatory frameworks of export-processing zones, where foreign investors can take advantage of reduced tax rates and other incentives for their companies (see Rivas, 2014, pp. 96–9).

The Salvadoran government actively takes part in the formation of transnational business relationships through the establishment of legal frameworks and institutions to attract investors. El Salvador becomes a brand, situated in strategic cultural, geographic, economic, and political (and, crucially, post-war) spaces. Along with the privatization of the telecommunications sector in 1998, the government opened its markets to foreign investors through the establishment of PROESA in June 2000.[7]

The agency's promotional material available on the Internet during those first years of existence presented an impressive, ambitious list of possibilities for diversifying the Salvadoran economy.[8]

"[PROESA] seeks to generate employment, transfer technology and aid the country's development process. The Agency's primary objective is to attract and assist foreign direct investment in industries such as: agro industry, textiles and apparel, contact centres, light manufacturing and electronics, logistics and distribution centres, software development, tourism and footwear ... PROESA has helped more than 138 multinational firms expand or establish operations in El Salvador. Successful companies have proved that El Salvador works for their business and that Salvadoran people work to make their firms even more productive and profitable."[9] The agency's logo and slogan, "El Salvador Works," represents a strategic link between labour, place, and bodies: "The slogan 'El Salvador Works' has a dual meaning: the people of El Salvador work hard, and El Salvador works for businesses. Warm colours were chosen for the logo. A circular shape that encloses three people united by the arms represents the human connection."[10]

Using the slogan "El Salvador Works" and a logo that "represents the human connection," the government agency promoted and marketed El Salvador as a "brand that travels around the world," with an appealing set of characteristics represented in the warm colours of the logo – shades of beige and brown, and earth tones. Scholars of advertising and visual culture have argued that advertisements do not sell a product; rather, they persuade, selling ideas and a lifestyle associated with the represented product: "advertising functions in a much more indirect way to sell lifestyle and identification with brand names and corporate logos" (Sturken and Cartwright, 2001, p. 198).

The human connection represented in the logo could persuade us of the desirable characteristics of Salvadoran workers. Only arms, among the most important "links" in outsourcing operations and global labour, are shown. These body parts represent a new and even more flexible iteration of the Bracero Program, in which the mobile Mexican labour force was recruited to the United States during the mid-twentieth century for agricultural work; it was so named because of the strength of their arms. PROESA's website also refers to Salvadoran labour at a regional and global level, about which the executive director states: "The human quality of our people makes our work force a very special one, so special that the Salvadoran labor force is recognized throughout the region and the world."[11] Readily available human capital, both within and outside El Salvador, is presented as one of the country's main assets and brand elements.

The reference to Salvadoran narratives of work ethic is important on this website, which primarily promotes the capacities of Salvadorans within the country. It is particularly crucial in a context where regional and global or transnational labour migration has historically been a common strategy for economic survival. Approximately 20 per cent of Salvadorans presently live and work outside El Salvador; the acknowledgment of its diaspora should further underscore the extent to which El Salvador is connected to globalization processes and is attractive to foreign investors. This tells us about the ideas that PROESA's investment promoters might have about the mobility of many Salvadorans, and how this mobility might, paradoxically, inform investment promotion strategies that depend on a stable, locally available, culturally informed, and skilled workforce in San Salvador.

Media institutions also play a significant role in the circulation of the discourse of investment and transnationalism, especially by publishing newspaper articles that highlight the novelty of bilingual call centres and the role of PROESA in promoting investment in El Salvador.[12] A December 2003 article in the high-circulation Salvadoran daily *La Prensa Gráfica*, titled "Call centers in search of Salvadoran voices: Two new centers in 2004,"[13] illustrates the presentation of the voices of Salvadorans – in this case, by investment promoters and the journalists who report on these events – as a coveted export commodity, emphasizing the flexibility of the Salvadoran worker and the strategic location of El Salvador as an ideal place for investment. As the executive director of PROESA explained to the reporters, with the new call centres "we will export voices from El Salvador."[14] Representatives of this governmental investment promotion agency noted the advantages that, in their view, El Salvador offered to companies interested in establishing call centre operations: "The modernization in telecommunications, the bilingual workforce, the [convenient] time zones, neutral accent and the distances between the country and North America have positioned El Salvador as one of the profitable nations for this business."[15] The main requirement for employment, as the article notes, is "perfect English" because the call centre employees "will wait on North American customers."[16] Through constructs of flexibility and bilingualism, El Salvador becomes a strategic location and part of a naturalized array of logics and hierarchical flows. Some histories and accents are made invisible, while certain language skills and labour flexibility seem prestigious and thus especially valued and marketed to potential investors.

Several of my interviews with managers at call centres illustrate and add nuance to these ideas and institutional discourses. Rodolfo is a recruitment manager at a call centre that provides technical support for a company based in the United States. During our interview Rodolfo spoke about the connections between global media technologies, Salvadoran migrants in the United States, the end of armed conflicts in the region, and the presence of call centres in El Salvador: "Since the war ended [in 1992] the country, the government, has tried to reach out to foreign countries to invest in El Salvador by having a series of plans, for example infrastructure. The telecommunications infrastructure in Central America, specifically El Salvador, is very, very modern, very advanced." Rodolfo continued, "On the other hand, due to the massive emigration of Salvadorans to the [United] States and through the years, and the cultural influence, culturally we are very, or up to a point, similar to the [United States], and there is much affinity with U.S. culture and U.S. society because of so many expatriates being from this country there."[17]

Rodolfo places the call centre and the Salvadoran investment climate within an interesting and complex historical, social, and political frame. "Since the war ended" is a historical moment linked to events that have affected the Salvadoran population, both in El Salvador and in the United States. This ethnoscape of Salvadorans creates and sustains transnational practices and connections between both countries. Clearly, the sometimes uneasy and contradictory flows of people, commodities, cultural expressions, and discursive forms and practices have converged – like undersea fibre-optic lines and other infrastructure – and become constitutive of post-war Salvadoran national and transnational identities. Cultural links are imagined, aided, and enmeshed in the new, real, and "advanced" infrastructure.[18] Salvadorans who live in the United States often continue to communicate and maintain meaningful ties with friends and family in El Salvador, shaping their perceptions of and affinities to both countries. A familiar and marketable linguistic, cultural, and historical narrative emerges in this way. Rodolfo readily adapts this market knowledge to his description of investment and the work of customer service – in addition to the crucial aspects of technology, Salvadoran agents in the call centres have a cultural knowledge that allows them to relate well to their North American customers. It is to this understanding and enactment of the marketable (and contradictory) characteristics of prospective call centre workers that I now turn.

Contradictions and Opportunities: "We Need People Who Speak English"

Prospective call centre employees in El Salvador undergo a series of interviews and assessments, including placement tests and language evaluations, to determine their competence in English, especially speaking and listening comprehension. The managers described how this process unfolded in their human resources departments. Many of these activities and evaluations seem to be standardized (or are at least quite similar) across the bilingual call centres that I researched in San Salvador, according to information shared by the interviewees about this topic.

During our interview Gabriela was enthusiastic and emphatic about her hiring goals for the new call centre. "We don't want even the second-best; we want the best talent of El Salvador. That is our goal, to become the first employment option, so everyone will want to work with us, [and] we need people who speak English."[19] Advertising is an important first element of reaching this goal. Once people respond to employment advertisements, a lengthy process begins in the call centre's offices in San Salvador. In some cases about one month of interviews, tests, and paid training workshops passes between the first interview and the first actual day of work. Not everybody who is initially screened is employed and given a headset. At the call centre where Rodolfo works, what he called the hiring "take rate" or "staffing yield" for some accounts is only about 8.2 per cent of the applicants who show up for the initial screening. To give me an idea of what this process entails, Gabriela explained the procedures of her workplace in detail:

> Gabriela: First we receive all the candidates interested in participating in the process [...] The first thing that is done is a telephone screening, to know the level of English. If they can maintain a conversation over the phone, that is the most basic thing, and there we ask them certain questions to know more about the candidate's level of English. After this, if the person fulfills the requirements, we invite him or her to take a computerized test to measure their skills, technical as well as in personality, which he or she must have according to the profile we need for the position. If the person passes – and this is something very interesting – because if the person passes this test, the next step is an interview with the managers, who are fully bilingual. But if they do not pass, we invite

them to take a course to improve, depending on how much they need it, and we give them a second chance, so they can retake the test.

CECILIA: And why is this, because there aren't enough bilingual people?

GABRIELA: No, the truth is that *[pauses, and briefly switches from Spanish to English] the lack of skills in terms of language* is so large that we need to invest in the people so they can master the language. First the language, and then improve the accent.[20]

Gabriela's explanation (and her brief yet notable code switching from Spanish to English) highlights an important point. Even as the process aims to be highly selective – involving aptitude tests and several interviews on the telephone and in person – the demand for employees is high enough that when not enough applicants are fully qualified for specific accounts, the call centre has to produce *exactly* the competent and fluent English speaker, the Salvadoran employee with an improved accent of its imaginary. Spoken English needs to be processed; the employer has to "invest in the people" to take advantage of the decision to invest its capital in El Salvador and fulfil the aspiration to be the "number one" employer. In other words, the ideal, bilingual Salvadoran call centre agent does not exist naturally as an immediately available workforce. This subject is shaped and produced by the very technologies of service and language that he or she ultimately serves.[21] As prospective call centre employees take language tests, employers are conducting surveillance and training in an effort to construct and evaluate competence and the linguistic capital represented by English language fluency.

Fluent, punctual, and trustworthy call centre agents are vital for these companies, which are new investors in the country, while the employees themselves are also new to the industry and to this type of job. In some ways this is a complementary relationship – and also a contradictory one. "The quality of our work depends on the people we are able to hire," explained Sara, a human resources manager whom I interviewed in 2006. It seems to me that this is among the most interesting and concise descriptions of the challenges of call centre operations in this location, an insight that further illuminates the contradictory aspects of English-language fluency in Salvadoran call centre hiring.

Sara had previous employment experience with non-governmental organizations before joining the human resources department of a Canadian-European company whose agents handle mail-in rebates and the collection of late bill payments, among other services, for North

American companies that have outsourced these business operations. At the time of my interview with Sara this company had been present in the Central American region for nearly a decade, first in Guatemala and then recently establishing call centre operations in El Salvador. They had plans to eventually branch out to Nicaragua and Panama. Clearly, they were in a very active expansion and hiring mode regionally. Everything was new – even the recently installed security system in Sara's office building. I interviewed Sara within her first three weeks of employment, which she described as extremely busy. New groups of approximately twenty-five agents were beginning their training every week. Impressively, Sara had hired fifty-three people so far, while she was in the process of negotiating with clients for possible new accounts and interviewing candidates for several positions in other areas of the company besides customer-service and call centre agents. She had recently processed the employment paperwork for an accountant and receptionists. Sara was already steeped in her work and enthusiastically recounted some of her experiences with evaluating language, screening, and interviewing prospective employees.

As we discussed these topics, Sara turned to her computer screen and shared a slide presentation with me, detailing the vision, history, and corporate strategy of the company. This lengthy slideshow was in all likelihood used to pitch the call centre services to clients in other markets and countries, and now to potential new clients in El Salvador. It listed the many regional partners and markets with which the company was associated, implying that the clients who decided to use the call centre services of this company could join this special group of recognizable brands. The themes of this slideshow were "sun" and "shore," creative representations of the ideas that the sun never sets in this company's call centres and that strategically located Central America is not just one more offshoring site; it is indeed the ideal and central place for these operations. The background image of most of the slides was a picture of a sun rising over clouds and the waters of an ocean. The design of this presentation is part of the company's advertising; it sends deliberate signals to the world of potential clients in this imaginary of post-war, rebuilt, and investment-ready Central America. The brand elements on display – round-the-clock hard work and accessibility – add to the discourses of other call centres, rival outsourcing operations, investment promoters, and the timely, appealing qualities of call centre operations in the isthmus. "When your sun is rising, ours is too. And when your day is over, ours keeps on going round the clock," Sara read from the presentation. She proceeded with the slideshow. "The time has come to outsource your business processing needs to the Best Shore, a new Central America where you'll find a friendly business environment, modern telecommunications, and an eager, highly skilled workforce that speaks both English and Spanish."[22]

The initial screening and hiring process at Sara's workplace is similar to that described by Gabriela: applicants present their résumés, interview with the managers, and take an English-language test to determine their eligibility. If the applicant passes the evaluations, the company then asks for references and a police record. Sara explained: "In this world, if you ask for a native speaker of the language, someone who speaks English [fluently], you can get just about anything. Many of the applicants we receive are either deported or are people who maybe had legal problems."[23]

Understandably, Sara takes these issues seriously. "Their English might be perfect. So the screening we do now is thorough. We ask for prison record, police record ... precisely this is not about finding out the bad things about people before knowing them. But [it is about] having standards that are more or less acceptable." She elaborated: "In this case we also must take care. We are hiring people to have interpersonal relationships with others. And maybe they are not responsible."[24]

I asked her if the specific fact of being a deportee from the United States meant that an applicant could not be hired. For Sara, this is not the primary element of the applicant's file that determines employment eligibility in El Salvador. Instead she considers the circumstances of deportation along with other relevant factors associated with each applicant's particular case. Some people have overstayed their tourist visas, and others have been deported after serving criminal sentences in the United States. She explains: "If they have been deported for any criminal act, there is nothing we can do. All of us need a second chance, and we have the will to help. But we cannot expose our other employees, or trust the agent with expensive equipment."[25]

The possible association of criminality with a fluent or native speaker of English – even if the Salvadoran deportee is not a criminal or a gang member – reveals another facet of the history of migration and the relationship between the United States, El Salvador, and the call centres. The presence of deportees in this workplace highlights the increased significance of migration, to the Salvadoran transnational and national imaginary and to notions such as the brand "El Salvador Works." Deportation and the social categories and subjects it produces become part of this reality. These social categories are not neat, and they create spaces of doubt and exclusion (see Coutin, 2010).

Where the applicant learned English marks the asymmetry between countries and the transnational inequalities between Salvadorans. The prospective employee who learned English at a bilingual school or an after-school language academy in San Salvador has a legible, local explanation for her or his linguistic capital. In contrast to the deportee, this local bilingual person does not have a possibly misread (or in some cases suspect) migration history to explain during the interview. In some sense the deportee is still living a migratory process and explaining her or his displacement. Speaking English "like a native," often perceived as a sign of integration into mainstream society in the United States, becomes entangled with a history of deportation from the United States and the socio-economic diversity of English speakers in El Salvador.

In the call centres, human resources managers are aware that applicants have learned English under different circumstances, while also holding the idea that the call centre can be an essentially equal workplace where fluency in this language can be an eraser of other differences and understandings of status. The deportee, an embodiment of displacement and the effects of migration, is stigmatized upon forced return to El Salvador and when job hunting in San Salvador's call centres. The cases of deportees working at call centres during this time seemed anecdotal, yet real and abundant; while I was conducting research in El Salvador, if the subject came up in informal conversations with call centre agents, they would usually answer something like, "Yes, of course we have entire teams made up of deportees" (Rivas, 2014, pp. 108–13).

Another interesting aspect of this dynamic of globalization and displacement is that although many women work at the call centres, it is primarily young men who draw attention and scrutiny as deportees and call centre workers (as the interview with Sara reveals). Reena Patel (2010) argues that women in India have re-spatialized both the call centres, where they work the night shift, and their routes to and from work. Owing to time-zone differences, women in Mumbai travel at night in streets that may often be unwelcoming to their presence. Patel engages categories such as domesticity, exploitation, danger, and liberation in women's work "as *spaces* that individuals embody and experience, often in overlapping and conflicting ways" (p. 11, emphasis original). In connection with this gendered angle on call centres and mobility – where Indian women predominate and face specific challenges in the

industry and beyond – the narrative and themes of young male deportees who may or may not be trustworthy or dangerous embody the
contradictions of call centre employment and transnational displacement in San Salvador. These young men are simultaneously representative and challenging of globalization's ideas of promise, anxiety, and
change.

While I was conducting these interviews, I noticed the circulation of
another image, a way to appeal to the interests of Salvadorans who lived
in North America and perhaps attract them to the call centres. The ideal-
ized call centre workers of the Salvadoran imaginary might look like the
young people in the "Meet Your Roots: Work in El Salvador" campaign.
"Meet Your Roots" is a program that was organized by PROESA in 2006
and is specifically aimed at second-generation Salvadorans living in the
United States and Canada who may want to learn about El Salvador.
The main idea of the program is that young Salvadoran Americans (or
Canadians of Salvadoran descent, if that is the case) come to El Salvador
from these countries and work at a call centre for a determined period
of time, usually one year. During that time they can travel within El
Salvador and learn about Salvadoran culture. "Meet Your Roots" is
presented as an opportunity to reconnect with the country from which
parents (and grandparents) emigrated, and to enhance the skills of these
young Salvadorans for future professional growth after their year in
El Salvador. PROESA explained in a March 2006 press release: "This
would increase the possibilities for young people that participate in the
program in North America since upon return they would be prepared
to assume employment challenges that require bilingual personnel,
opportunities that tend to increase owing to the growth of the Latin[o]
community in the United States and Canada."[26]

An advertisement accompanies this press release. "El Salvador's
Government and the Call Center Industry in El Salvador invite you
to rediscover a totally new country, ideal for living and working," it
states.[27] Who would not want to employ the young people featured in
this advertisement, in San Salvador's call centres and beyond? They
are young, presumably fluent in English, and completely familiar with
the United States – but not in a "dangerous" way, because they do not
appear to be gang-affiliated deportees. They are friendly, ready with
their headsets and a smile, and dressed in a professional yet relaxed
manner. Most important, they *want* to live in El Salvador, to work, to
reconnect with Salvadoran culture and national identity, and to have
fun on their days off, as the smaller, inset pictures of sunny beaches and

a bar suggest. The advertisement extends an invitation to the daughters and sons of Salvadorans in the diaspora (many of them of rural or urban working-class origins) to reincorporate in the nation, to return and be productive, and to become part of a new reality of labour that seems far removed from the "roots" of their parents. The advertisement for "Meet Your Roots" represents the aspiration of ideal transnationalism, of young professionals who cross borders freely, who find unburdened possibilities in North America and in Central America.

The investment promotion agency wants to "export voices" and market the "roots" and bilingualism of its national and transnational workforce. Meanwhile, in call centre human resources departments such as Sara's and Gabriela's, transnational labour takes more complex forms – especially as these young managers face the real applicants, listen to the stories behind their English, try to fulfil the demands of hiring, and manage the everyday limitations and anxieties that surround the origins of the much-desired linguistic capital of the ideal "native speaker" call centre employee.

Culture on the Telephone

Bilingualism and flexibility are important and commodified qualities for call centre operations in San Salvador. The reach of English as the language of global corporations seeps into the views of competitive and professional behaviour. To attract global capital, the national investment promotion project depends on the presence and availability of a competent subject who can function in a bilingual environment – primarily, someone who is fluent in English.

Call centre work is represented as a desirable job option for English-speaking Salvadorans because it involves the use of linguistic and technical skills that are valued in a global economy. Spanish-English bilingualism is an important asset for current and prospective employees in this sector, and training often emphasizes listening comprehension and the improvement of vocabulary, particularly in competent performance of an American English accent.

Like their bilingual or English-speaking counterparts, Salvadoran agents in Spanish-only call centres are presented in PROESA's brochures and website as speakers with objective neutral accents. According to this claim, Salvadoran Spanish is not as defined or recognizable as are the distinctive intonations from other parts of Latin America. This assumption of neutrality – in speaking English or Spanish – is part of the brand presented in "El Salvador Works." The PROESA website claims that "Spanish spoken by Salvadorans is characterized by a neutral accent, another point in favor of the industry, opening up the possibility of answering calls from many different Spanish-speaking countries at one single place."[28]

However, Spanish is far from neutral. Anthropologist Pedro Geoffroy Rivas has argued that Salvadoran Spanish, identified in his 1978 linguistic study as *lengua salvadoreña* (Salvadoran tongue or language), is the unique product of centuries of colonialism, followed by nearly two hundred years of independent life (see Geoffroy Rivas, 1978). Salvadoran Spanish is a result of this regional and global history – of indigenous languages, colonization, conflict and coexistence with other Central American countries, immigration from different regions of Spain and other parts of the Spanish-speaking world, contact with the United States, and other changes that emerge from people's creative uses of language (see Geoffroy Rivas, 1979). The language spoken by Salvadorans is unfixed, shifting, and innovative, and it is intertwined with social, political, and historical practices.

The idea of a neutrality of language in El Salvador's present glosses over the racial and class origins of many of its speakers. The idea of Spanish-language neutrality in contemporary El Salvador echoes a long-standing idea of the cultural homogeneity of the Salvadoran population. The discourse of *mestizaje* as an unproblematic or harmonious mixture of people, essential to the coherent narrative of the modern nation, is often simplified in this way: El Salvador's indigenous population effectively "vanished," adopted the use of the Spanish language, and dropped its own Nahuatl language as well as the *refajo* – the traditionally hand-woven dress of women and girls – and other cultural norms (persuasive critiques of this view include Peterson, 2007). This "vanishing" was due in part to the violence against indigenous populations, historically and especially after the 1932 Matanza, when up to thirty thousand people – indigenous inhabitants, peasants, and presumed participants in a labour uprising near the coffee plantations of western El Salvador – were killed over several weeks (Lindo-Fuentes, Ching, and Lara-Martínez, 2007). Historians and other scholars of this event point to the fact that the memories and narratives of 1932 have served as a foundation for various explanations of racial, political, and class attitudes and for hegemonic understandings of El Salvador as a modern "mestizo nation" that does not pay much interest or attention to its indigenous roots.

A manager whom I interviewed in 2005 astutely noted that, while learning English was currently essential for people who applied for jobs at call centres in El Salvador, "How many people speak Nahuatl or are even interested in learning it? We're very strange in that way ... El Salvador absorbs things from outside easily, but does not care about its own identity." Clearly, English and Spanish bilingualism is more valued and prioritized in the logic of the call centres and the post-war Salvadoran economy. As portrayed in PROESA's campaign, for young Salvadorans in the diaspora, meeting their roots means reconnecting to a modern, globally linked country and to call centre employment (and not, in this case, to indigenous languages). John Lipski discusses the differences in language usage that arise from inequalities in Salvadoran society, especially in the area of access to education, observing "a sociolinguistic discontinuity between the speech of marginalized groups and that of the urban middle and professional class. The latter, increasingly, turns outward to Mexico, Spain, and the United States for advanced training, acquiring in the process a de-regionalized language, while the socially marginalized sectors advance in their linguistic evolution at an ever greater rate. The discrepancy is immediately noticed upon listening to a Salvadorian, whose socioeconomic origins can be identified after only a few words" (Lipski, 1994, p. 256).

The neutral Spanish and English (with an improved accent) that are supposedly spoken in call centres, then, aim to be global and more prestigious forms, to be "de-regionalized" and professional national languages that are readily marketable to the global consumer. In the context of language and commercialization Arlene Dávila discusses the creation and portrayal "of a neutral or universal version of Hispanidad – the putatively neutral, 'non-accented' Spanish and 'generic' Latin look" (Dávila, 2001, p. 93). She argues that the representation of Latinos in the United States by advertising agencies is overwhelmed by the codes and symbols of middle- or upper-class, fully bilingual, university-educated Latin Americans who arrive in the United States as adults to work as marketing creatives. The investment strategy involved in marketing and promoting the notion of linguistically de-regionalized Salvadoran call centre agents is part of post-war economic development projects, in which Salvadoran business sectors and the government are able to present the linguistic abilities of Salvadoran workers as neutral, flexible, and fit for a global market. The marginal, poor, or speaker of indigenous languages is not visible in this imaginary.

The focus on marketing neutral Spanish might lead one to believe that Spanish-language call centres attend exclusively to callers from other Latin American countries or to Spanish-speaking callers from the United States. Many of the calls are international, but these call centres also serve Salvadoran companies and callers in need of local customer service. For the managers of these centres, it is crucial to maintain this local and regional perspective. "Each market is a different culture on the telephone," said Elizabeth, speaking of her work at a Spanish-language call centre.[29] She meant that, for each client account (for example, an airline or a telephone service), the call centre workers had to get to know well a brand (and its customers). In addition, agents had to become familiar with slang or specific words depending on the countries and regions from which the calls were routed – not because they would use this slang themselves during the conversation, but because it would be easier to understand the usage of customers.

Conclusion

The call centre, where voices are trained and commercialized for export, is an example of the many reconfigurations of the national and the transnational as experienced by Salvadorans in these emerging workplaces and institutions. The search for an ideal neutral accent in Spanish and for a native speaker of English involves a larger process of surveillance and regulation, as anxieties about the return of bodies from the United States conflict with the economic need to commercialize their language skills.

This work shapes and is shaped by the Salvadoran context, part of a post-war moment in this imaginary, where personal aspirations for mobility are caught up in projects of national mobility – for instance, in the desire for prestigious foreign investment. There is tension in this imaginary: a brand such as "El Salvador Works" emphasizes the special characteristics of the *local* Salvadoran workforce, while also projecting itself as globally appealing. Migration is an important element of this brand and this imaginary, to the point of trying to attract young Salvadoran Americans in search of their roots, while trying to negotiate the categories of deportation and possible criminality.

Salvadoran call centre agents are hired as bilingual, fluent, and deterritorialized voices. Production is segmented, outsourced, and divided dramatically between the American (or Americanized) consumer, in this case imagined as a monolingual English speaker, and the bilingual

Salvadoran agent. English proficiency is commercialized and tied to opportunities for career advancement, as portrayed by recruiters, investors, newspapers, and advertisements. In this way the Salvadoran voice as a commodity is branded, exported, and projected to the global economy.[30]

NOTES

1 Interview with Laura and Elizabeth, 16 August 2005.
2 Interviews with call centre employees were conducted in confidentiality, and I refer to them in forms that are appropriate to this context. All names are pseudonyms. I conducted most interviews in Spanish and have translated these (and other documents and newspaper articles, when necessary) into English.
3 "'Call centers' en busca de voces salvadoreñas: Dos nuevos centros en el 2004," *La Prensa Gráfica*, 4 December 2003, 41.
4 M.G. Arun, "Last call: India's about to hang up on call centre culture," *India Today*, 15 March 2013, http://indiatoday.intoday.in/story/bpo-industry-call-centre-culture-dying-in-india/1/258032.html (accessed 17 November 2015).
5 Amanda Rodas, "Call center crean 1750 plazas a pesar de pobre oferta bilingüe," *El Diario de Hoy*, 3 November 2012, http://www.elsalvador.com/mwedh/nota/nota_completa.asp?idCat=47654&idArt=7399253 (accessed 18 April 2013).
6 Guadalupe Hernández, "Convergys busca convertirse en el mejor 'call center' para trabajar," *El Diario de Hoy*, 31 March 2014, http://www.elsalvador.com/mwedh/nota/nota_expansion.asp?idCat=75293&idArt=8664641 (accessed 30 January 2015).
7 PROESA was created in the year 2000. However, by executive decree on 17 May 2011 the agency was re-created as Agencia de Promoción de Exportaciones e Inversiones de El Salvador, with the following objectives: "promotion of exports of goods and services produced in El Salvador," "promotion and attraction of foreign investment," and "promotion and strategic direction of the collaborative frameworks of investment between the public sector and the private sector." ("Decreto No. 59, Creación de la Agencia de Promoción de Exportaciones e Inversiones de El Salvador," *Diario Oficial* (San Salvador, El Salvador), 391, no. 90, 17 May 2011, 8–11.)
8 The PROESA web pages to which I refer in this chapter include the following, initially consulted in 2006 and 2007: "PROESA: Contact Centers," http://www.proesa.com.sv/industry/contact_centers.htm (accessed 29 October 2006); "PROESA: El Salvador Works; A Brand That Travels around

the World," http://www.proesa.com.sv/about/brand.htm (accessed 3 April 2007); "PROESA: Strategic Locations," http://www.proesa.com.sv/why/strategic_location.htm (accessed 30 October 2006); and "PROESA: Telecommunications," http://www.proesa.com.sv/infrastructure/telecommunications.htm (accessed 31 October 2006). The website for this agency has changed in format and modified its content over the years and under different administrations, but consistently it has sections that provide information similar to that on the sites I initially consulted, detailing El Salvador's investment climate, past and current foreign investors, and other economic indicators. PROESA's more recent web pages include http://inversiones.proesa.gob.sv/ (accessed 6 February 2013) and http://www.proesa.gob.sv/ (accessed 18 June 2013).

9 "PROESA: Who We Are," http://www.proesa.com.sv (accessed 3 April 2007).

10 "PROESA: El Salvador Works; A Brand That Travels around the World," http://www.proesa.com.sv/about/brand.htm (accessed 3 April 2007).

11 "PROESA: Strategic Locations," http://www.proesa.com.sv/why/strategic_location.htm (accessed 30 October 2006).

12 During 2003, PROESA facilitated the establishment of twenty-three foreign companies in various sectors, including textiles and clothing, electronics, tourism, and call centres. From 2000 (when PROESA was created) to 2003, 106 companies invested in El Salvador, creating thirty thousand jobs. The clothing sector was predominant during these years, with 22,835 jobs created. At that time 1,500 jobs were in call centres. (Luis Andrés Marroquín, "El año viejo deja 23 nuevas empresas: PROESA logra inversión con la marca El Salvador Works," *La Prensa Gráfica*, 23 December 2003, 72.)

13 "'Call centers' en busca de voces salvadoreñas: Dos nuevos centros en el 2004," *La Prensa Gráfica*, 4 December 2003, 41.

14 Ibid.

15 Ibid.

16 Ibid.

17 Interview with Rodolfo, 6 September 2005.

18 During 2005 and 2006 a fibre-optic network covering approximately 475 kilometres was installed in El Salvador, to link it to other countries of the region. (Alexander Torres, "Navega.com invertirá $20 mill [million] en red fibra óptica," *La Prensa Gráfica*, 5 May 2006, 64–5.)

19 Interview with Gabriela, 23 August 2005.

20 Ibid.

21 Along with these processes that were taking place at the call centres, various Salvadoran institutions recognized the importance of and demand for

English speakers in this job and the investment landscape. The National English Center (NEC) was formed during this time to address the demands for qualified candidates for call centre employment. It emerged from the collaboration between the Salvadoran ministry of education, PROESA, and other organizations. In 2006 the NEC was administered by ITCA (an educational and technological institute) and FEPADE (an educational development foundation). The NEC offered "specialized English courses for the call centers field" (advertisement for NEC, *La Prensa Gráfica*, 8 May 2006, Oferta Empleos, 13). The first group of NEC graduates completed their ten-week course in July and August 2006 (Reina María Aguilar, "Primera promoción del Centro Nacional de Inglés," *La Prensa Gráfica*, 3 August 2006, 54; and Clara Villatoro, "Nuevos jóvenes bilingües para 'call centers,'" *La Prensa Gráfica*, 22 July 2006, 24).

22 Interview with Sara, 15 August 2006.
23 Ibid.
24 Ibid.
25 Ibid.
26 PROESA, "Meet Your Roots Is Launched in Dallas," press release, 18 March 2006, http://www.proesa.com.sv/news/2006/press270306_1.htm (accessed 2 April 2007).
27 Ibid.
28 "PROESA: Contact Centers," http://www.proesa.com.sv/industry/contact_centers.htm (accessed 29 October 2006).
29 Interview with Laura and Elizabeth, 16 August 2005.
30 Portions of this chapter were previously published in the book *Salvadoran Imaginaries: Mediated Identities and Cultures of Consumption* (Rutgers University Press, 2014).

REFERENCES

Appadurai, A. (1996). *Modernity at large: Cultural dimensions of globalization.* Minneapolis, MN: University of Minnesota Press.
Ball, K. (2003). Categorizing the workers: Electronic surveillance and social ordering in the call center. In D. Lyon (Ed.), *Surveillance as social sorting: Privacy, risk, and digital discrimination* (pp. 201–25). London, UK: Routledge.
Coutin, S.B. (2010). Exiled by law: Deportation and the inviability of life. In N. De Genova & N. Peutz (Eds.), *The deportation regime: Sovereignty, space, and the freedom of movement* (pp. 351–70). Durham, NC: Duke University Press. http://dx.doi.org/10.1215/9780822391340-014

Dávila, A. (2001). *Latinos, Inc.: The marketing and making of a people.* Berkeley, CA: University of California Press.

Geoffroy Rivas, P. (1978). *La lengua salvadoreña.* San Salvador, El Salvador: Ministerio de Educación, Dirección de Publicaciones.

Geoffroy Rivas, P. (1979). *El español que hablamos en El Salvador* (4th ed.). San Salvador, El Salvador: Ministerio de Educación, Dirección de Publicaciones.

Lindo-Fuentes, H., Ching, E., & Lara-Martínez, R.A. (2007). *Remembering a massacre in El Salvador: The insurrection of 1932, Roque Dalton, and the politics of historical memory.* Albuquerque, NM: University of New Mexico Press.

Lipski, J.M. (1994). *Latin American Spanish.* London, UK: Longman.

Patel, R. (2010). *Working the night shift: Women in India's call center industry.* Stanford, CA: Stanford University Press.

Peterson, B.G. (2007). Remains out of place: Race, trauma, and nationalism in El Salvador. *Anthropological Theory, 7*(1), 59–77. http://dx.doi.org/10.1177/1463499607074293

Rivas, C.M. (2014). *Salvadoran imaginaries: Mediated identities and cultures of consumption.* New Brunswick, NJ: Rutgers University Press.

Sturken, M., and Cartwright, L. (2001). *Practices of looking: An introduction to visual culture.* Oxford, UK: Oxford University Press.

3 Growing Downhill? Contestations of Sovereignty and the Creation of Itinerant Workers in Guyanese Call Centres

ALISSA TROTZ, KIRAN MIRCHANDANI,
AND IMAN KHAN

Introduction

In the global clamour to attract foreign capital Guyana is a small player with high stakes. As in many other countries that house transnational call centres, the enticement of foreign capital has been an important part of state economic policy for the past three decades. Alongside its attempt to generate capital flows into the country, however, Guyana is unique in the scale of its simultaneous outflow of educated citizens in response to few local employment prospects and an ambivalent state policy that has both lamented outmigration and fostered the reliance on remittances from overseas Guyanese. This was the backdrop and condition of possibility for the entry of the transnational call centre sector to Guyana less than a decade ago. Explicitly constructed as a shift away from the routine factory work that global manufacturing firms provide, telemediated customer service work is frequently promoted as creating white-collar jobs for educated youth. Such official optimism is belied by the wry observation of Ray, a young man whom we interviewed for this project, who noted that "the call centre seems to be a growing factor in Guyana, even though its growing seems to be going on a downhill."

In this chapter we trace the practices through which Guyana's transnational call centres are constructed as sites of hope and progress by state, organizational, and media actors. We explore three forms of sovereignty that are continually being enacted and expressed within Guyanese call centres – state sovereignty, customer sovereignty, and the sovereignty of global capital. We draw on interviews with seventeen call centre workers to explore their experiences in the sector, as well as two key informants who reflect on state and organizational policies relating to the emergence of the call centre industry in Guyana. Our findings reveal that, despite the promise of high-paying, stable jobs, call centres are highly itinerant spaces, with employees bearing the brunt of firm closures and contractions. Instead of providing sustainable careers for educated youth, jobs are designed to be low paid and extractive. With striking uniformity the workers we interviewed noted that call centre jobs cast Guyana as a nation with a young, captive workforce ripe for exploitation by foreign actors.

Contested Sovereignty and Guyanese National Identity

The contested nature of the sovereignty of modern states is now well established. Traditional constructions of states as holding unilateral, complete, and final decision-making authority (referred to broadly as the Westphalian international order) over their territories and citizens have been overtaken by the recognition that states are one of multiple actors that exercise sovereignty within national spaces, with varying degrees of success and coherence (for an excellent overview see Trouillot, 2001). While some theorists argue that the capacity of nation states to influence local and international policy has been eclipsed by the reach of global capital (see Hardt & Negri, 2000), others suggest that state sovereignty has not so much declined as it has been transformed, taking, as Aihwa Ong notes, a "graduated" form (Ong, 2007; also see Brown, 2010; Sassen, 1996). State practices collude with global capital and have differing effects on various groups of citizens depending on their deemed "usefulness" to processes of capital accumulation. In this context, states play a key role in mediating the tensions between global and national interests in relation to market ideologies (Brown, 2010). The project of constructing the "wall" along the U.S. southern border, for example, is a state strategy to negotiate the tensions between the needs of North American businesses for the cheap labour of illegal migrants, and the demands of workers on the U.S. side who are affected by the downward pressure on wages and working conditions brought about by easy access to this labour (Brown, 2010). At the same time, the sovereignty of states and of global capital is continually contested through the elaboration of what Lisa Lowe refers to as a transnational counterpolitics (Lowe, 2008), as well as through the emergence of global cosmopolitan norms (Benhabib, 2007) that aim to universalize rather than segment the access to livelihoods and rights. These sovereignties are exercised in conjunction with another extremely significant actor having sovereign power in the call centre setting – the customer. Goldberg (2009) has observed that the customers' exercise of power is privatized as an expression of "preference" that can supersede state or global actors in the context of neoliberal, market-oriented global capitalism. In service encounters, for example, customers are made to feel sovereign in two ways – first, by being positioned as "relationally superior" to the server on the other end of the call and, second, through the construction of service exchanges as ones in which customers are "in charge" (Korczynski, 2007, p. 78). Although this power may be far from complete, customers' real or imagined preferences play an important role in the ways in which states exercise their sovereignty. As discussed in the sections to follow, the transnational call centre in Guyana is one site where these varied sovereignties materialize and overlap and are visibly in contest.

Methodology

The analysis in this chapter is based on reviews of local newspapers, company descriptions, promotional materials, and websites that refer to Guyana's call centre industries. These include several YouTube video recordings of site visits and interviews with company officials. In addition, seventeen call centre workers (none in a managerial position) and two key informants were anonymously interviewed, based on contacts made through sources that were unconnected to call centre administrators. Interviews were held in a home space after working hours, and workers were invited to share their job experiences without employer involvement or knowledge. Pseudonyms have been used, and specific information, particularly about employers and work situations that might identify individuals, has been omitted to protect worker identities.

Companies solicit applications in a variety of ways, placing advertisements in newspapers and on television as well as using social media tools like Facebook to advertise recruitment drives and provide information to prospective applicants. Responses on social media represented a significant way to gauge someone's comfort level with using computers and the Internet, all important attributes for a job in the industry. A survey of advertisements, websites, and Facebook pages reveals an emphasis on attracting applications from those with educational qualifications (in several cases, evidence of the successful completion of secondary-school-leaving examinations) and communication skills. In return, potential recruits are "promised" excellent remuneration and a highly professional and transnational work environment that entails handling overseas clients (one can interact with the outside while remaining in the country). The Facebook page of one company carries an image of a cheque with the payee section reading "Pay yourself GYD$250,000 (USD$1,207)." This is no small amount in a country where the minimum monthly wage is GY$35,000 (US$169). Facebook pages also regularly feature company promotions for employees, like competitions for trips to Kaieteur Falls (a local tourist destination in Guyana's hinterland) and to neighbouring islands like Trinidad and Tobago. Pictures of company-sponsored outings – picnics, sports days – for employees and their families are also posted, and team loyalty is encouraged through participation in tournaments (dominoes, basketball, and cricket) that pit company employees against other teams around the city and country. The larger companies in particular are also involved in various activities that confirm their narratives of being good corporate citizens. Employees volunteer in several company-sponsored initiatives that range from donating food and other goods to orphanages and senior citizen homes, to participating in clean-up activities in the capital city, Georgetown.

Of the seventeen people interviewed, twelve were currently employed in a call centre, and the remaining had recently left call centre jobs. Fifteen of the respondents were women, and the average age of the interviewees was twenty-two years, reflecting the official emphasis on providing employment for youth. All the people interviewed had completed secondary education, and thirteen had gone on to further studies, with two people also receiving some of their tertiary education in other Caribbean countries before returning to Guyana. Such educational diversity raises important questions about job opportunities as well as the possibilities for career fulfilment within the call service sector.

Exercising State Sovereignty by Selling Guyana

In the mid-1980s Guyana began a process of economic and political liberalization, after a decade and a half of a failed cooperative socialist experiment in the immediate post-independence era, embarking upon a home-grown version of structural adjustment in 1985 and ushering in democratic elections in 1992, the first since 1964.[1] Under the Economic Recovery Program (as it is locally referred to, and which has been continued or intensified by the administration that has been in power since 1992), this process has included aggressively courting private capital in such areas as logging, ecotourism, mining, and oil exploration;

pursuing public-private partnerships; and putting mechanisms in place to attract and facilitate foreign investment, like GO-Invest, a governmental organization that offers tax breaks, duty-free concessions, and other perks to would-be investors. Such initiatives signalled a shift towards the "graduated sovereignty" of the Guyanese state whereby "corporations [hold] an indirect power over the political conditions of citizens in zones that are differently articulated to global production and financial circuits" (Ong, 2007, p. 78).

While the bulk of the investment opportunities so far have been in the country's hinterland (home to most of Guyana's indigenous peoples, and the site of diamonds and other precious metals and one of the largest remaining rainforests), one of the key demonstrations of interest in investing on the coast (where some 90 per cent of Guyana's population resides) has come from call centres and business-processing outsourcing firms (BPOs). Emerging on the local scene over the past ten or so years, this is still a relatively incipient market, providing employment to about three thousand people, most of them young women between the ages of seventeen and twenty-five years. With about a dozen call centres in operation, the sector is dominated by one large firm that has locations also in Southeast Asia and Latin America, and which grew from a workforce of less than one hundred people a decade ago to being reportedly the largest private employer in the country today with over a thousand employees. There are a few medium-sized enterprises with over a hundred workers (these appear mainly to be companies that have established themselves only in Guyana), and some much smaller and struggling firms. Most of the companies are located in or near the main urban centre and capital city of Georgetown. According to one media report, they "perform such services as outbound sales / telemarketing; inbound customer support; voicemail transcription; medical records transcription; and data warehousing for clients in the United Kingdom, the United States, Canada and Mexico" (*Guyana Chronicle*, 3 July 2012).[2]

Effusive predictions by state and industry officials that "Guyana's future as an outsourcing destination is brightening by the day" (Ammachchi, n.d.) are accompanied by news reports of expansion, such as an announcement that a global call centre organization is currently in the process of building what is described as one of the world's largest call centres in Guyana, a multi-purpose site that will include educational, interfaith, and other facilities for the six thousand additional employees whom it hopes to attract (*Guyana Chronicle*,

5 September 2013). In addition, Teleperformance USA is reported to have invested US$2 billion to set up a 1,500-seat call centre that is anticipated to be operational by 2016 (*Stabroek News*, 2015). Call service investment is described by the government of Guyana not only as essential to economic diversification but specifically as creating high-value jobs in information and communications technology,[3] with official efforts to improve infrastructure and put in place regulatory frameworks to support increased investment in this sector.[4]

This emphasis on creating highly skilled service sector jobs and specific assurances that call centres offer what one company describes as the "ultimate dream career" assumes particular urgency in a country from which some 43 per cent of secondary-educated citizens and 89 per cent of tertiary-educated citizens migrate to OECD countries, the highest rates in the world (Mishra, 2006; Thomas, 2014b). Although the migration rates appear to have slowed over the last two decades, they remain comparatively high (Vezzoli, 2014). Guyana is one of the few countries in the world that has been experiencing a negative net-population-growth rate since the 1980s. Preliminary results from the 2012 Population and Housing Census suggest that the downward trend continues, with a recorded drop in the population from 751,223 in 2002 to 747,884 at the 2012 count, which analysts attribute to the continuing and unsustainable external migration (Guyana Population and Housing Census, 2012; Thomas, 2014a, 2014b). The scale of the outflow was precipitated by the economic crisis facing the country in the 1970s, and the continuing attrition suggests that the majority of educated Guyanese still see their future as lying outside of the country, which is perhaps not surprising given some official estimates that youth unemployment was as high as 42 per cent in 2012.[5] In accounting for Guyana's emergence in the last decade as a desirable destination, it is not insignificant that this demographic is targeted by the call centre industry and constructed by the state as providing a key national edge in the global competition for jobs. As our interviews with workers revealed, Guyana's comparative advantage appears to lie in the supply of a low-waged, professional, and captive workforce.

Moreover, the global geography of the contemporary outsourced call centre industry is significantly structured by histories of colonialism. Not coincidently, the countries colonized by Britain and the United States (such as India and the Philippines), currently dominate the industry. Guyana's stated "promise" as a potential site for such work is situated similarly in its colonial past, and histories of exploitation, destruction, and extraction are celebrated in glossy promotional material as a "heritage" that lends itself perfectly to the needs of a burgeoning new industry, with the accompanying and essential benefit of a widespread use of standard English.

In fact, it is "location, location, location" (in the words of one company) that is singled out as Guyana's strategic advantage. In a highly competitive global environment, state and industry officials promote Guyana, a relative newcomer to the call centre scene, as an "alternative India" and a "complement to the Philippines."[6] Location here has multiple meanings. First, to the North American market in particular, the country is represented as one that offers physical proximity. A mere six-hour flight away from the United States and Canada and in the same time zone (an hour ahead during daylight saving time in North America), Guyana is sold in some advertisements as an ideal "nearshore" site, especially for companies thinking of venturing into outsourcing for the first time. As one local website boasts, "operating nearshore enables [X] to offer world-class service and quality at a fraction of the cost, compared to US call centers."

In addition, Guyana is home to a highly literate and educated English-speaking population that is located on the northwestern shoulder of South America, bordered by Venezuela, Brazil, and Suriname. The country was colonized first by the Dutch, and then by the British, who took control in the late-nineteenth century, unifying the three Dutch colonies of Demerara, Berbice, and Essequibo into British Guiana in 1831. The only English-speaking country in South America, Guyana has historically been seen – and sees itself – as fundamentally Caribbean. At the level of interstate relations, Guyana is a founding member of the Caribbean Community and Common Market, CARICOM. In more popular terms, the country has consistently supplied team members to the West Indies cricket team, for years a symbol of Caribbean unity and resistance. However, in recent years there has been increasing interest in pursuing bilateral relations with neighbouring Latin American countries as well as regional initiatives with South America. Extra-regional migration from the country has been historically to the United Kingdom and since the early 1960s to the United States and Canada; within the region it is primarily to the anglophone Caribbean, although there is also regular movement across the country's borders to Suriname, French Guiana, Brazil, and Venezuela. Against this backdrop, location is meant to signal cultural and linguistic overlap and proximity with the predominantly North American client base of the companies, therefore promising an excellent return on investment. In one promotional video, scenes from around the country are accompanied by a voiceover that describes Guyana as "an excellent opportunity for the world outsourcing market, because [of] English, native language English, high scalability and the cost structure, so at the end of the day, it is that alternative to India, it is that complement to the Philippines."[7] As the chief executive officer of one company remarked, "[our] largest market is serving customers in the United States, and Guyana understands Americans. They speak the same native-English language and watch the same TV" (*Stabroek Daily News*, 3 September 2013).

Cost competitiveness includes the fact that there seems to be relatively little outlay on the intensive accent training that takes place in other locations (see Mirchandani, 2012). We came across only one company that mentioned training on "speech and diction, Americanization, accent training," unlike call centres in India, in which language and accent training are universal. In fact, the people to whom we spoke seemed surprised to learn that such emphasis was devoted in other countries to inculcating a certain way of speaking English, and they surmised that – in addition to familiarity with North America via migratory histories (remittances constitute one of the major sources of the country's foreign exchange earnings), saturation via television programming, the Internet, and the circulation of commodities (in this regard the country is similar to other call centre locations) – Guyanese spoke English with what was described as a flatter, more internationally comprehensible accent. Several workers whom we interviewed stated that they believed Guyana to be an ideal location for outsourced customer service work because of its linguistic advantage.[8] Stephanie reports that customer compliments are often based on perceptions of her superior accent compared to that of call centre workers from other countries. She elaborates on this principle of distinction: "The person speaking to you may not be able to assist you ... they have other persons from Spanish-speaking countries, so their English wouldn't be as good as mine ... so they would be like, 'I was talking to someone and I couldn't understand them ... your English is so much clearer,' ... As much as we are in South America ... we don't speak Spanish, so that's a good thing for us." Similarly Troy reports: "A lot of my customers would be like, 'Where are you from?' I say, 'Guyana. That's in South America.' And they say, 'Wow. Your English is perfect. Your English is better than mine!'" We were told in one interview that pass rates for call centre English tests ranged regularly in the eightieth percentile and higher, as opposed to pass rates in Latin American countries of between 10 and 25 per cent. Moreover, respondents not only attempted to distinguish themselves from workers in non-English-speaking countries but also compared the Guyanese accent favourably with the accent of more popular offshore call centre sites like India and the Philippines, as well as other English-speaking Caribbean countries like Jamaica and Barbados that have a far longer history of continuous and direct interaction with "foreigners" through investment and highly developed tourism industries. Interviewees distinguished the Guyanese accent as being far less heavy and much easier to understand; in fact, Belize was the only other country that was singled out as having a relatively neutral English accent. As we shall see later, however, workers report that, while some customers compliment them on their accents, others do not believe that Guyanese English is "superior" to the English spoken in other call centres.

In the context of Guyana, the call centre sector not only occupies a unique economic niche in so far as it promises to provide accessible, prestigious, and professional jobs for the segment of the population that is most prone to migration from the country, but is also an important symbol of Guyana's future prospects for social and economic wealth, as well as a marker of the country's proximity to the West. To pursue its neoliberal agenda, and much like other Caribbean countries over the last two decades, the Guyanese state has developed incentives, institutional sites, and promotional material to clearly send the signal to foreign investors that it is open for business (Klak & Myers, 1998), while also engaging locally in a comprehensive branding campaign to promote the idea that the infusion of global capital is in the country's national interest. Do employee narratives resonate with the anticipated convergence of foreign capital and local development promised by these official representations? Odessa, a young woman in her mid-twenties, stated that Guyana as a nation benefits from the industry: "not only are they gaining some sort of revenue for the country, they are also providing jobs for persons who need the jobs, and they are also giving you skill in [the] customer service line." Her optimism, however, was tempered by the sense that the benefits from the call centre industry are unevenly distributed within the country, a sentiment shared by all the young people we interviewed for this study.

Such scepticism reflects a wider suspicion about government policy that seems unable to stem the persistently high emigration of educated citizens. According to one of our key informants, the distrust is also fuelled by a lack of transparency: agreements made between foreign investors and the government are not public knowledge. Shantie noted that she would like to stay in Guyana because she likes living close to family but knows that she might have to leave if she is unable to find suitable employment related to her interests and education. She explained: "A lot of Guyanese have less faith in the government for providing jobs for them. Everybody is trying to do their own thing – open a shop here, open sewing, try to do nails. You lose hope in your government ... I think people in high places are more concerned about their welfare and the welfare of their families. The government body is not making decisions to benefit the citizens as a whole."

Drawing extensively on our interviews with present and past call centre employees, the remainder of this chapter attempts to make sense of this significant discrepancy between the highly optimistic tone represented by the state and the call centre industry and the overwhelming frustration expressed by those who have been named as local beneficiaries.

The "Guyanese Wage" and the Sovereignty of Global Capital

A prominent transnational call centre in Guyana recently announced plans to set up a new centre that would be "amongst the largest contact centers in the world." Officials of the company provide several reasons for its decision to make "its largest investment to date" in Guyana, including the country's location, English proficiency, and state support. The announcement was also welcomed by Guyana's president, who noted, "The Government of Guyana recognizes and appreciates the work being done by [the company] in providing jobs for hundreds of our people."[9] A company video also highlighted the contributions that it would make to the local economy through the provision of over six thousand jobs and the construction of an energy-efficient building that would include water harvesting, solar energy, and state-of-the-art facilities for workers.[10]

Despite this rhetoric about the possible benefits accruing to Guyana, such as potential revenue, employment, and customer service training, all the respondents noted that Guyana's primary competitive advantage lay in the unavailability of jobs for young, educated people, which was reflected in the staggering emigration levels of those with secondary schooling and beyond, precisely the group interviewed for this project. The workers we interviewed had all sought call centre jobs after having completed their post-secondary education in such subjects as communications, social work, business management, computer studies, industrial relations, accounting, and tourism. Ray, who had recently left his job as a call centre operator in the hope of finding something more suitable to his training, had sent out more than fifty job applications to no avail and was on the verge of losing his apartment; he observed: "Job opportunity is lacking. You know, there isn't much for young people to do in Guyana … I now see why it is young people turn to guns, young people turn to drugs, young people sell their bodies to make money. Because you have no other choice. People have qualifications [but] are being paid next to nothing."

This sense of having little to no choice (a situation that leaves one vulnerable to exploitation, with one interviewee using terms like *guinea pigs*, *test rabbits*, and *lab rats* to convey her sense that Guyanese employees have little negotiating power) overwhelmingly framed workers' descriptions of a highly unequal playing field tilted in favour of companies. Despite recruitment advertisements claiming that call centre jobs provided wages comparable to those of professional jobs (such as accountants or mid-level managers) in Guyana, no one whom we interviewed indicated that the job was well paid or a match for the training they had received, with most describing call centre work as something they turned to because they were unable to find employment in their areas, and as "better than no job." Marlia, who had a diploma in business management, as well as secretarial qualifications, said that she felt overqualified for her position at the call centre, but "I wanted a job. I didn't want to be home doing nothing, and you're waiting and waiting."

This bleak employment scenario translates into a context in which, having few competitors, employers can set low wage levels and minimal benefits and yet ensure a continuous supply of prospective workers. Cynthia, who worked for a call centre for more than five years, stated that even if they provided the best service, they could not earn more than what was deemed appropriate for a "third-world worker," what she called "a Guyanese wage."

Salzinger (2003), in her ethnography of workers in Mexican maquiladoras, argues that firms create rather than simply find the workers whom they hire. Like the women who produce televisions on assembly lines in Mexico, Guyanese call centre workers are subject to hypersurveillance and poor wages. Paying a "Guyanese wage" to workers with few other employment options fosters a sense of both entrapment and docility because their jobs are unsustainable and simultaneously play a key role in maintaining the precarious balance of inputs that workers receive from a variety of sources to piece together their liveli-hoods. The greatest challenge mentioned by most respondents was the fact that, despite having professional, full-time jobs, they were unable to make enough to cover their rent and food bills. As Marisa noted, "the main challenge for me is the money. [We] just go and work there because [we] are not getting anything else. But that's a huge problem." Respondents observed that their jobs seemed to call for a well-educated workforce, but the levels of remuneration and the working conditions were established not in the light of living costs but in terms of the low-est possible wage that would ensure a continuous supply of labour. In the context of a nation in which educated youth have severely lim-ited work options, this wage is extremely depressed. About half of the respondents had worked in at least two call centres, and the monthly take-home pay ranged from a low of GY$30,000 (US$144) to a high of GY$50,000 (US$241), with over two-thirds earning between GY$40,000 and GY$45,000 (US$193–US$217). Three people indicated that they had been promised much higher monthly salaries at one company (roughly GY$70,000, or US$338), but these had not materialized, and two of the interviewees recalled one company that had for a short while only offered a significantly higher salary than had the other centres.

The average monthly earnings of those employed at the time of the interviews was GY$44,000 (US$212), not significantly higher (especially given the educational qualifications of our respondents) than the current legal minimum wage of GY$35,000 (US$169). They do not begin to cover the daily expenses of running a household, given the recommended basket of goods for a family of four (two adults and two children), esti-mated at GY$59,972 (US$289) in December 2013; notably, the basket was restricted to food items and did not include housing, health care, trans-portation, and utilities (*Stabroek Daily News*, 30 March 2014). Shantie, who had received approximately GY$40,000 (US$193) per month, but had been laid off from her job, said, "Who can live on 40,000 when the month come ... because my rent alone is 40,000 and you have to pay transpor-tation and food." Indeed, even local managers seemed to recognize the disconnect between wages and the cost of living; a company video that focused on efforts being undertaken to reduce absenteeism among its workforce identified the provision of coffee and crackers as important incentives for employees to come to work regularly and on time. [11]

To make ends meet, workers must rely on a combination of wages, performance incentives, and company handouts, as well as income from other sources. Base salaries are complemented by incentives that workers can receive for meeting or exceeding their targets. Agents in Guyana are more likely to receive supermarket vouchers than the leisure-related luxury items that many call centre organizations provide as incentives.[12] Workers said that they preferred vouchers because vouchers were untaxed, and they depended on them to meet their basic needs. Companies offer extremely low wages in conjunction with corporate "perks," which include things like competitions for teamwork or milk programs for women with children, but many of these handouts are "earned" through performance and are not automatically given. While organizations construct these incentives in order to motivate employees and to transform "productivity pursuits into an exercise of 'fun'" (Alferoff & Knights, 2003, p. 76), as in the case of interviewees who reported to us that their employer sometimes held competitions for things like "best hair day" and "best dressed day," workers report that they rely on bonuses to meet their food needs and to make ends meet. As one woman pointedly told us, "I am sure most persons would forego the incentives for more money on their pay cheques."

Therefore, despite having full-time jobs, many workers were frequently forced to turn to family or friends at home and abroad for help or to seek supplemental self-employment incomes. Marlia pointed out that "they have some people who have more than one job ... when they come off, they go and do a receptionist job or part-time work somewhere ... The money couldn't do." Cindy said, "The salary that I am working for can't even look after me and my daughter. If it wasn't for my child's father, I don't know what would happen." Lisa described the salary she received as "harsh" but noted that her company provided some flexibility to its employees and had schedules that allowed them to attend school if needed, "especially if you don't have responsibilities and you're still living with mom and dad, it's nice." Sherry commented similarly: "it's kind of like a temporary place when you're done school. It's not something you can support your family on. It's not something you can have two kids and pay your rent with. Or actually support a household."

Call centres in Guyana are predominantly staffed by women workers. Korvajärvi (2009), in her ethnography in Finland, traces three ways in which call centre work is feminized. First, work is numerically dominated by women. Second, interactive service work is seen to be closely related to the traditionally feminized roles of serving and caring. Finally, the structural conditions of work such as the low pay and the limited possibility of career advantage are seen to be appropriate for women (Korvajärvi, 2009, p. 132). For Guyanese women, however, as elsewhere in the Caribbean, there is also a strong relationship between motherhood and waged work, which makes them more reliable employees and more likely to seek call centre jobs. As Lisa explained, "Women are coming forward, women are coming forward, men in Guyana ... I call [Guyanese men] 'not working millionaires.' But women, when you see you have responsibilities, you wanna earn a dollar."

Workers therefore confront the contradictions of being involved in feminized work that is "professional," yet poorly paid and demeaning. In their research on call centres in India, D'Cruz and Noronha (2006) trace the ways in which professionalism is "at the heart of organizational control" in customer service work (p. 344). While traditional notions of professionalism relate to employee control and discretion over work, a call centre worker is "defined as a person who has the desire to satisfy customers, put aside personal problems and concentrates on service, accepts stringent monitoring and self-timings, is able to withstand strains and pressures of work, and is receptive to the idea of taking on another identity in the interest of the organization and the customer" (p. 346). Indeed, employees report little control and discretion over their timings and scripts. Marlia noted, "You don't get a break. You are only allowed a half-hour lunch. You might get a chance to go [to the] washroom, but as soon as you come off a call, there is another one waiting for you."

Van den Broek (2002) describes surveillance systems typical of call centre work environments around the world, and the centres in Guyana clearly follow these geographical blueprints. Work environments consist of open-plan offices, large display boards on which the productivity statistics for each employee are prominently visible and continuously updated, daily reports on average handling times and numbers of calls, and automated telephone dialling systems that minimize wait times between calls (pp. 49–50). Methods of public shaming for unmet quotas are also common. Shantie described a "big board" at her call centre that was updated every fifteen minutes and could be seen by all agents and supervisors; she felt embarrassed by the publicly displayed low scores that she sometimes received, which had her "depressed." Shannon shared her experience of a call in which the customer had been abusive, and her supervisors had felt that she had been rude. She was asked to write "lines." Clearly angry and humiliated by this disciplinary strategy, Shannon noted, "I will never forget this thing – a yellow paper saying 'I will not interrupt the customer or speak rudely to the customer.' I had to do back and front. And I told them. 'I am getting paid for this? I am not writing lines! I am not in school! I never write lines in school and I'm not going to write it now.' So then I got disciplinary action. I didn't get suspended, but I got a warning letter."

Employees are also exposed to industry instability. According to the workers interviewed, hiring and firing occurs continuously at call centres (there was a strong sense among interviewees that one "got through right away" with a job there because of the "high turnover"), even though one manager to whom we spoke insisted that there was a very low turnover rate of employees. Ray pointed out, "In my opinion it's one of the easiest jobs to get ... so because of that a lot of people are going into call centre work ... [it] has been turning over employment all the time ... You see someone, 'Oh, you're working at a call centre,' and next couple of mornings you're no longer there." Despite the stressful nature of their work and the low pay they receive, workers report being treated as easily replaceable. Sherry recounted a situation in which her manager called her in to the office to inform her that she was being fired. She was told that one of her calls had been randomly surveyed and found to be unacceptable and that she had also been absent from work for one day. Sherry informed the manager that she had called the supervisor because she was sick, and she wanted more information on the unacceptable call, but she received no further details. Ria had a similar experience: "I was given extra work to do that I was not paid for, so I spoke up about that, and it was a problem ... I stood up and said, 'It's unfair for you guys to give me extra work and I'm not being paid for it.' So that was a big issue, and they made an excuse that I was rude to a customer, which I was not. And that was my reason for being fired."

Workers' experiences reveal the significant volatility built into outsourced customer service work, particularly in some of the smaller firms that rely heavily on outgoing calls in which agents make sales pitches for various products and services, and are promised incentives depending on the number of "leads" they score per shift.[13] "Campaigns," sometimes only months long, are held by companies that then hire workers to complete the task. Companies may downsize drastically and rapidly with the loss of contracts, and workers bear the brunt of these shifts. One company, for example, offered jobs to workers but, after several months, revealed that it was not able to pay wages. Shantie was an extremely successful employee who earned three times her base wage in performance incentives one month. She received an award for being the best agent. However, with the loss of the contract, she and her colleagues were fired without being paid for completed months of work. To no avail, she made several telephone calls, asking to be paid for the work she had completed. Instead, she received an email from a manager who accused her of being disrespectful, even as the company admitted it was simply unable to find monies to pay employees. Shantie noted the futility of making a complaint to the state, pointing to her sense that the sovereignty of the state had been eclipsed by the sovereignty of global capital, and that her firm held special status that allowed for such blatant evasions of local labour laws. She went on to discuss her feelings of exploitation in terms related to national identities, and said of her past employers: "[Foreigners], they kind of don't like Guyanese people. So I felt disrespected. Like, 'Do you think you [foreigners] could come to Guyana and exploit Guyanese people?' Even though my passport is Guyanese, I still know about human rights. I still have a foreign mentality. I was so mad that they thought they could come to Guyana and wuk [work] everybody out and have everybody ... like little guinea pigs or test rabbits, or lab rats, and 'you just working on us. You know, and you guys are not paying.'"

Although Shantie anticipated that making a complaint to the state would not benefit her, a group of workers in Linden, a mining town with only one call centre, was reported to have successfully filed such a complaint in September 2015. Workers complained to the Ministry of Social Protection because they had lost their jobs without prior notice. The company was deemed to have violated local labour laws and was asked to compensate workers (*Kaeiteur News*, 2015) although enforcement remains uncertain.

Even when companies follow local labour laws, the wages do not allow sustainable livelihoods. The "Guyanese wage" has both an empirical and an ideological dimension. Empirically, it is a wage that is lower than the one that companies would have to pay to workers doing equivalent work in the United States, and it is not linked to the actual cost of living in Guyana. Ideologically, it is a wage paid to workers who have no other alternatives and are conceptualized as people who should be grateful for receiving anything. Bearing the cost of unpaid wages is transferred to workers whose proper place is seen as one of respect for managers and corporations even if the business owners' attempts to generate contracts for Guyana-based companies fail.

Sovereign Customers and Abusive Consequences

In addition to the pace of work and continuous monitoring, one of the most stressful dimensions of working in a call centre is the irateness of many callers. Korczynski (2007) notes that service work is "structured by the dual and potentially contradictory logics of bureaucratization and customer orientation" (p. 79). The logic of bureaucratization prompts organizations to rationalize processes and streamline work in order to reduce costs and increase efficiency. At the same time, customer orientation requires a flexible and resource-rich approach to satisfy the broad range of customer needs. Korczynski and Ott (2004) argue that management reconciles the dual logics of customer orientation and organizational efficiency by promoting a "myth of customer sovereignty": "this form of enchantment is promoted such that within the interaction it appears to the customer that he/she is sovereign while at the same time creating space for the frontline worker to guide the customer through the constraints of production" (p. 581). While such sovereignty may be mythical from the perspective of customers, for call centre workers the customer responses to their service have a significant and direct impact on their conditions of work as well as their wages and prospects for future employment. Customers exercise sovereignty in two ways: first, through the exercise of the power that arises from their role as the "served" during customer-service interactions and, second, through their regional location in capital-rich countries that are being served by citizens of formerly colonized, poorer countries, which are positioned as labour providers in the global economy.

In the course of telephone conversations with customers asking for information or technical help or to place orders, workers do the emotional labour simultaneously of being empathetic, anticipating and attempting to meet customer needs, and of managing their own feelings. In addition, they must do all of this while meeting their daily targets. Shannon offered an example: "In the midst of the conversation someone would be like, 'Oh, you know what happened to me today … my dog died …,' but you can't laugh … you gotta be like, 'I'm so sorry to hear that … How old was your dog?' That can also make the call overlap the time … you can get penalized [for that] … if you're too empathetic, they penalize you …'Oh, this agent was too empathetic; she could have been empathetic but still made the call shorter.'"

In addition to being empathetic, managing customer anger is seen to be part of all customer service jobs. In transnational customer service exchanges, however, calls are frequently infused with enactments of racialized hierarchies. Despite Guyana's often-cited locational advantage in terms of language, as mentioned earlier, workers in Guyana note that hierarchies of race are often expressed in terms of the focus on the deficiencies of Guyanese accents. Sherry commented: "When some of the customers pick up on the accent, depending on how you speak to them, when they pick up on the accent and see that it's an outsourced centre, they give you a hell of a time and say, 'I can't understand you. You're not speaking English.'" Accents are, however, also heard in terms of the places to which speakers belong. Shannon shared an experience in which she was disciplined for being rude to a customer who "was cursing and cursing. I was like, 'Ma-am, if you keep cursing, I'm going to have to disconnect the call.' [The customer said,] 'You don't tell me what to effing do.' And I warned her again. Then she calmed down. Then she was like, 'You know what, I don't know why I call these bullshit places.' And then I got really rude. I said, 'Ma-am, you need to listen. If you don't listen, you won't get anywhere. You need to shut up and listen.' … My quality analyst was listening … and I have to go into the training room and get disciplinary action."

As Shannon's account reveals, customers who make derogatory comments about the "bullshit" places they are calling are the most difficult to deal with. Marissa noted: "Sometimes they just hang up. As soon as they realize you are not American, they just hang up. 'Where am I calling today?' 'You've reached Guyana.' They just hang up. And sometimes they say, 'Why is it that a call centre that is American has to be all the way in South America, you know, outsourcing, and Americans need jobs?'" As a result of such interactions, workers are continually required to manage their emotions during calls. Odessa described her response to abusive callers: "I roll my eyes. I sit back. I breathe. Then I start to speak. I say, alright. Then I go into this tone. And I say xyz and try to see what the problem is. You have to relax and let them vent for a couple of minutes. And then they would be like, 'Are you there?' And I say, 'Yes, I'm here. I'm just listening to you.' Then they come down because they just want to get that anger out of their system."

This emotional labour serves to get customers into the correct "mood" so that they can be served by call centre workers. Ahmed (2014) notes that "mood work" is an important part of the construction of nations. She argues that national moods, for example the collective feelings of Americans against outsourcing, serve to identify and name "moody figures" who are assumed to "get in the way of attunement" (p. 14). Mood work is done not only by customers who express anger on the telephone, and therefore assert their citizenship right to outsourced jobs, but also by customer service agents like Odessa who, by allowing callers to "get that anger out," attempt to transform their moods.

Customer anger about outsourced jobs is often expressed in terms of objections to accents. While workers are often told by customers that their accents are deficient, as we discussed earlier there is no evidence of the sort of widespread training that takes place in Indian call centres and which focuses on accent "neutralization." On the contrary, Guyana's locational advantage is seen to lie in the language complementarity between workers and customers. As Cowie (2007) has shown, the notion of neutral English that is valorized in customer service work promotes the ideal of a nationless English that is understood by all. In practice, notions of appropriate and inappropriate accents infuse call centre work, and the idea of a "neutral, globally understood" English is in fact code for English spoken by a particular class of people, from particular regions. Most of the training received by workers in Guyana, however, was basic and geared towards using the databases in order to provide information, belying the official promotional material that promises highly skilled and well-trained opportunities. As Julie observed, "The key skills are just computer knowledge ... say, product searching, you should know how to navigate to work in a timely manner to assist the customer. And the way you speak. I guess that's basically it, just the computer knowledge and the way you speak so they could understand you." Although basic grammatical tests were administered at some of the sites, no one mentioned receiving the intensive voice and accent training that is common in call centres in other parts of the world. For Shannon, "when you do the first training, there is a list of words that everyone has to go through and become familiar with ... but they don't constantly check you or come back and say, 'Oh, this agent said a word wrong,' or 'This agent don't pronounce this word correctly' ... If Miami [i.e., offshore quality-control analysts who listen randomly to calls] is listening, they might give a warning." Joanne remembered that she was simply asked to "pronounce certain words ... words that customers would often use or words that you would have to use with customers." Audrey recalled being explicitly advised that it would be easier to acquire a British accent in order to be understood, but this was not a job requirement (employees could use computer programs to practise pronouncing words, but this did not appear to be formally integrated into a consistent training program). Marlia described the training she received: "They would just train you – communication – how to talk, how to address a call, customer issues. For example, if [the customers] call and they [didn't] get their phone, well, they gotta do this, that's how they work you, how they train you. They don't train you on how to talk to a customer. If a customer curses you, there is a script. There is everything in the system. The foreign world, they don't play to curse you out [customers don't hesitate to verbally abuse you at the slightest opportunity]. That's the foreign world. But you have to pick up the script and tell them. But you can't tell them anything without picking up that script."

The English demanded in these transnational service interactions, then, is hegemonic and not universally understood. Parvati pointed out that "they have an accent to me and I have an accent to them … but if I can speak to you and you can understand me, then we can get the transaction done." The workers may struggle at times, as Audrey observes: "The most important thing is to listen to the customer and understand exactly what the customer is saying … For instance, we are calling England, and most of us don't understand the English people when they talk very fast." But the onus is on the local employees to learn to understand the customers and to adapt their own way of speaking to ensure that they are understood. In the absence of comprehensive training, they must do this largely on their own and incur the consequences of failing to "speak properly." Customer rage is significant not only because of the toll it takes on employees required to deal with a continuous stream of abuse, but also because it has a direct relationship to workers' remuneration. Latoya noted, "You could get a low score for sounding too foreign and not understanding the accent." During training, workers are told to expect hostility from those at the other end of the call and are taught strategies to defuse this anger. They are encouraged to be empathetic and understanding of customer frustrations and to deal with abusive customers in a calm and detached manner. Susie commented: "Most of the customers are arrogant, and they weren't listening to me. Some of them, I get them to calm down, and they would listen to me. But some of them are arrogant." For Marlia, customer feedback can result in an "autofail" or an immediate dismissal for the day: "Customers make you go home because when you call the customer back, they say, 'I don't like this person and I'm giving them 1.' If you give them 1, it's a low score. They pull you off the phone."

Even in the face of racially specific and sexually explicit harassment, workers are required to educate customers on proper telephone etiquette rather than disconnect the calls. Shannon shared her encounter of a customer who said, "Hi. Guess what I have today? … I have 8 inches." After three warnings she was able to finally disconnect this call. During her training, "they warned you about that. They warned you about a particular lady called the screamer who just used to call in and scream at you. No matter what you are doing or why, she just used to scream. Then they have this old perverted … customer who used to call every month … [and another] who used to want to know the colour of your breasts, if you're black, brown, what colour they are, what size you wear." Despite these abusive exercises of power, customers give scores to workers, and these scores are treated as decontextualized aggregate measures of performance.

Although workers face possible job termination for speaking back to customers, Shantie offered an instance of when she did so: "I think my first day I was cursed out. Someone said, 'Oh, fuck off.' … I am the type of person I get really offended based on how people speak to me … I think one guy in particular, he was impressed with my persistence because he was trying to insult my job. And I said, 'Not because you don't like my job means it's a degrading job. I probably caught you in a bad mood and I apologize for that, but I am not ashamed of my job in any way and I don't think any job should be frowned upon."

Other responses to unsatisfactory working conditions range from occasionally talking back to difficult customers, to finding ways of relieving stress on the job, as in the case of one young man who muted his telephone regularly in the face of abuse once he had figured out that he had a three-second span in which to do so. When he came back on the line, he would say, "I'm sorry about that, I had to cough" (or sneeze or something like that). Alternatively, people simply left the job, even if this was a risky decision to take given the bleak job prospects.

These individualized responses were the only recourse that employees appeared to have; without exception and despite numerous examples given to us of clear violations of Guyana's labour laws, none of the call centre workers expected to receive any official recourse on their behalf. For instance, during a service call that necessitated her calling another department to assist a customer, Marlia discovered that, unlike agents in other countries employed by the same company, Guyanese workers were required to work on Easter. Her colleague informed her that their local branch had staged a strike and no longer worked on holidays. Marlia reflected that no such protest could occur in Guyana because of the large supply of unemployed educated youth. As in other parts of the world, the industry is completely non-unionized, and so far there appear to be no efforts to organize workers or for employees to seek out a union; as Parvati observed, "it seems as though these people are so timid ... they are fearful they would wake up and hear they don't have a job." In one case, workers had been laid off without being paid; some had received weeks of training, and others had worked while being continuously promised that they would soon receive their salaries. We were told that several people had gone to look for other employment or had kept up hope that they would receive their monies. A few had brought the matter to the attention of the relevant government officers and ministries, but to no avail. As one of the key informants concluded, "young people really have nowhere to go. We are like a milking cow. People's rights can be and are violated on a daily basis." Interviewees repeatedly emphasized that when it came to the call centre industry, it seemed to them that the government was more interested in protecting the interests of employers than in protecting the rights of Guyanese citizens to a living wage. While some recognized the global structural asymmetries at work that gave overseas companies the upper hand, there was an overall sense that this was compounded by a deliberate set of national policies that favoured local elites and foreign capital. Responding to a question about who benefited from call centres, Shantie said: "For Guyana, the people in the high places like the government and the people the government allows to come into the country to do business ... The government body is not making decisions to benefit the citizens as a whole ... In school we are taught that a government is a body that is supposed to make laws and rules that would help the citizens; they are not doing that ... everything is messed up, they are investing money where they are reaping all the benefits and gains ... and the workers ... the labourers are being taken for granted."

Given these experiences, none of the current and past employees that we interviewed saw call centres as providing career opportunities for young people. Unlike the women in the Barbados informatics industry interviewed by anthropologist Carla Freeman (2000) in the 1990s, the Guyanese workers do not see themselves as pink-collar employees who actively invest in narratives of class differentiation based upon the work they do. Cynthia reported: "I kinda enjoy customer service, which is why I am there. But I don't think it's a career thing ... basically, well in Guyana, [on] the whole, you have so much qualifications and you still have a hard time to get a job. Call centre is easier. At the end of the day it doesn't pay a lot, but it's still a job. It's better than just sitting at home."

This is not to say that they do not articulate aspirations to social mobility based on their educational accomplishments. As Shantie reflected, "I would say that a good job is one that could sustain yourself ... you want to buy a car, put down on a house ... if you would get a job that could enable you to put down on a car or further your studies or accomplish some of your goals." However, the call centre was not seen as even approximating the interviewees' idea of a "good job"; it was described as temporary, something to be done until a better prospect opened up, a place where some fortunate individuals might get promotions that mattered, but where the majority of the workforce faced systematic exploitation with little recourse for protection of their labour rights. Respondents explicitly described the call centre as a "job of last resort." As Marlia explained, "I left [the company] because I felt like a slave. When I say slave, you were not getting ... for instance ... weekends ... I would have to be at work. They give you one day off a week, and you can't choose ... You're like a slave, working ... Salary is also small, and the people knock you off." Several workers told us that their family and friends did not see what they were doing as a prestigious occupation but as a low-status job, and a few stated that they were even embarrassed to admit to working there. Marlia reported that she was afraid to even tell people that she was working at a call centre. Marisa said: "My family, when they heard that I was working at a call centre ... everyone was disappointed ... because they said, 'Man, you're going to UG [University of Guyana] and you have your little CXCs and good grades' ... so they didn't think it was the best, but for me it's better than not having a job at all, so for that reason I decided to stay on until something better comes."

Despite having to manage others' disappointment, workers recognize that their choices are limited. Cindy said, "To be honest with you, [my family] don't like it ... but you know as I say ... you have to work ... I have to get money ... Their view for me is to get a government job." In fact, workers reported the widely held perception that call centres not only provided routinized and low-paying jobs but also exacted a significant emotional toll. As Shannon summarizes, "if someone hears you're working at [a call centre], they say, 'Wow, you got patience.'"

Conclusion

The call centre industry in Guyana represents a site where multiple sovereignties are manifested and contested. Expressions of state sovereignty, customer sovereignty, and global-capital sovereignty shape the experiences of call centre workers in the industry. In Guyana it would appear that the call centre industry is here to stay for the time being, with the country being identified by both the Guyana government and prospective and existing investors as an excellent location and site for rapid and massive expansion. It is also promoted as a solution to the dilemma facing school leavers, with many high-paying and secure jobs that offer upward mobility and bright futures for career-driven young people. Behind the glossy advertisements, however, lies a story that unfortunately appears to provide little evidence of benefits for local workers. In Guyana this materializes against a backdrop in which school leavers and graduating students with some level of tertiary education find few jobs to match their skills, and where unsustainably high levels of emigration attest to the overwhelming sense that young professionals can only fulfil their aspirations by leaving. Those who remain have few choices, leading to a situation in which call centres set the rules. Growth, then, is one sided, as Rick presciently observed: "it's growing because it's a job easy to get in, but I don't think [it's growing] effectively." Parvati summed up the situation: "I don't think Guyana benefits on the whole because it's outsourcing whenever it's done ... and it's mostly the investors and whoever is at the top who earn the most ... The agent has to do the work and they are just earning a piece of it ... Like they say, Guyana is a third-world country ... Even though we would provide quality customer service, they are not gonna pay us ... These foreign [businesses], coming in and exploiting people at home. I am thinking, as much as they are the advocates for the customer ... to make sure the customer is treated a certain way, ... at the end, who is my advocate? ... There must be somebody who is representing Guyana." Call centres are sold as a win-win deal for everyone: foreign and local capital, customers, the state, and the Guyanese people. In the final analysis, however, and for the young workers upon whose emotional labour the success of the industry ultimately depends, there is an overwhelming sense that the promissory note they have been issued bears simply the imprint of their exploitation. These expressions of discontent represent cracks in the assumed popular support that underlies the forms of sovereign control at play.

NOTES

1 A former British colony, Guyana became independent in 1966.
2 The vast majority of clients are in the United States.
3 Foreign exchange earnings have traditionally rested on three primary export sectors – bauxite, rice, and sugar – although, given the size of the Guyanese diaspora, remittances could be added to this list (*Kaieteur News*, 5 September 2013).
4 Telecommunications Bill 18 of 2012 – which addresses the liberalization of telecommunications – remains before a parliamentary select committee. With regard to some of the structural issues with which Guyana is grappling in its move to embrace ICT, a September 2014 report found that of the twenty-seven Caribbean countries surveyed, Guyana had the second-lowest Internet download speeds (after Cuba) http://www.ict-pulse.com/2014/09/snapshot-actual-internet-download-speeds-caribbean-september-2014/ (accessed 1 October 2014).
5 The youth unemployment figures are 46.9 per cent for women and 39.7 per cent for men. See the World Bank Report, http://data.worldbank.org/indicator/SL.UEM.1524.ZS (accessed 29 September 2014). Up-to-date and reliable data in Guyana are extremely difficult to find. There have been no recent labour force surveys, and so far preliminary census results include only the most basic demographic information.
6 See for example the promotional video at http://www.youtube.com/watch?v=VM2xIQ2KRfo) (accessed 25 September 2014).
7 Ibid.
8 See, for instance, "Guyana Leads Caribbean as a Call Centre Facility," *Guyana Inc.*, issue 6.
9 Quoted in a company news report: http://www.qualfon.com/blog/2013/09/04/september-4-2013-georgetown-guyana-qualfon-building-one-of-the-largest-contact-center-campuses-in-the-world-in-georgetown-guyana-and-creating-6000-new-jobs/ (accessed 25 September 2014).
10 https://www.youtube.com/watch?v=sl_rDtXtGUo (accessed 29 September 2014).
11 https://www.youtube.com/watch?v=QJgb5MXrD_U (accessed 29 September 2014).
12 In just two instances it was reported that high-performing employees could occasionally spin a wheel to earn a prize, usually a small electronic or durable household item like a toaster.
13 One of the larger firms relied primarily on incoming calls, with which there seemed to be less uncertainty and change in relation to its overseas clients.

REFERENCES

Ahmed, S.. (2014). Not in the mood. *New Formations: A journal of culture/theory/politics*, 82, 13–28.

Alferoff, C., & Knights, D. (2003). We're all partying here: Target and games, or targets as games in call centre management. In A. Carr & P. Hancock (Eds.), *Art and aesthetics at work* (pp. 70–90). Basingstoke, UK: Palgrave Macmillan.

Ammachchi, N. (n.d.). Guyana outsourcing: Positioned for growth. (Nearshore Americas Blog). Retrieved from http://www.nearshoreamericas.com/guyanas-outsourcing-sector-brighter-future/ (accessed 29 September 2014).

Benhabib, S. (2007). Twilight of sovereignty or the emergence of cosmopolitan norms? Rethinking citizenship in volatile times. *Citizenship Studies*, 11(1), 19–36. http://dx.doi.org/10.1080/13621020601099807

Brown, W. (2010). *Walled states, waning sovereignty.* New York, NY: Zone Books.

Cowie, C. (2007). The accents of outsourcing: the meanings of "neutral" in the Indian call centre industry. *World Englishes*, 26(3), 316–30. http://dx.doi.org/10.1111/j.1467-971X.2007.00511.x

D'Cruz, P., & Noronha, E. (2006). Being professional: Organizational control in Indian call centers. *Social Science Computer Review*, 24(3), 342–61. http://dx.doi.org/10.1177/0894439306287979

Freeman, C. (2000). *High-tech and high heels in the global economy: Women, work and pink collar identities in the Caribbean.* Chicago, IL: Duke University Press. http://dx.doi.org/10.1215/9780822380290

Goldberg, D.T. (2009). *The threat of race: Reflections on racial neoliberalism.* Oxford, UK: Blackwell Publishing.

Guyana Population and Housing Census. (2012). *Preliminary report: Call centres burgeoning, July 3.* Guyana Chronicle.

Guyana Chronicle. (2012, 3 July). Call Centres burgeoning. Retrieved from: http://guyanachronicle.com/call-centres-burgeoning/ (accessed 29 September 2014).

Guyana Chronicle. (2013, 5 September). Towards more employment… Qualfon turns Providence sod for 6,000 jobs centre – set to be one of largest call centres in the world. Retrieved from http://guyanachronicle.com/towards-more-employment-qualfon-turns-providence-sod-for-6000-jobs-centre-set-to-be-one-of-largest-call-centres-in-the-world/ (accessed 29 September 2014).

Hardt, M., & Negri, A. (2000). *Empire.* Cambridge, MA: Harvard University Press.

Kaieteur News. (2013, 5 September). Qualfon Guyana begins another multi-million dollar investment. Retrieved from http://www.kaieteurnewsonline.com/2013/09/05/qualfon-guyana-begins-another-multi-million-dollar-investment/ (accessed 29 September 2014).

Kaieteur News. (2015, 4 September). 90 lose jobs as Linden Call Centre closes. Retrieved from http://www.kaieteurnewsonline.com/2015/09/04/90-lose-jobs-as-linden-call-centre-closes/ (accessed 29 September 2014).

Klak, T., & Myers, G. (1998). How states sell their countries and their people. In T. Klak (Ed.), *Globalization and neoliberalism: The Caribbean context* (pp. 87–109). Lanham, MD: Rowman & Littlefield.

Korczynski, M. (2007). Understanding the contradictory lived experience of service work: The customer-oriented bureaucracy. In M. Korczynski & C.L. Macdonald (Eds.), *Service work: Critical perspectives* (pp. 73–90). Routledge.

Korczynski, M., & Ott, U. (2004). When production and consumption meet: Cultural contradictions and the enchanting myth of customer sovereignty. *Journal of Management Studies, 41*(4), 575–99. http://dx.doi.org/10.1111/j.1467-6486.2004.00445.x

Korvajärvi, P. (2009). Attracting customers through practising gender in call-centre work. *Work Organisation, Labour & Globalisation, 3*(1), 131–43. Retrieved from http://www.jstor.org/stable/10.13169/workorgalaboglob.3.1.0131 doi:1

Lowe, L. (2008). The gender of sovereignty. *The Scholar and Feminist Online, 6*(3).

Mirchandani, K. (2012). *Phone clones: Authenticity work in the transnational service economy*. Ithaca, NY: Cornell University Press.

Mishra, P. (2006). Emigration and brain drain: Evidence from the Caribbean. *IMF Working Paper 06/25*. Washington, DC: IMF.

Ong, A. (2007). *Neoliberalism as Exception: Mutations in citizenship and sovereignty*. Durham, NC: Duke University Press.

Salzinger, L. (2003). *Gender in production: Making workers in Mexico's global factories*. Berkeley, CA: University of California Press.

Sassen, S. (1996). *Losing control? Sovereignty in an age of globalization*. New York, NY: Columbia University Press.

Stabroek Daily News. (2013, 3 September). Qualfon's new call centre to have 3,500 seats. Retrieved from http://www.stabroeknews.com/2013/news/stories/09/03/qualfons-new-call-centre-to-have-3500-seats/ (accessed 25 September 2014).

Stabroek Daily News. (2014, 30 March). Pro-poor policies. Retrieved from http://www.stabroeknews.com/2014/news/stories/03/30/pro-poor-policies/ (accessed 25 September 2014).

Stabroek News. (2015, 14 October). Teleperformance call center operational. Retrieved from http://www.stabroeknews.com/2015/news/stories/10/14/teleperformance-call-centre-operational/ (accessed 4 November 2015).

Thomas, C. (2014a, 3 August). Guyana's recent population bombshell. *Stabroek Daily News* (accessed 29 September 2014).

Thomas, C.Y. (2014b, 10 August). Guyana's 2012 population census: Runaway brain drain rules! *Stabroek DailyNews*. Retrieved from http://www.stabroeknews.com/2014/features/08/10/guyanas-2012-population-census-runaway-brain-drain-rules/ (accessed 29 September 2014).

Trouillot, M. (2001). The anthropology of the state in the age of globalization: Close encounters of the deceptive kind. *Current Anthropology, 42*(1), 125–38. http://dx.doi.org/10.1086/318437

Van den Broek, D. (2002). Monitoring and surveillance in call centres: Some responses from Australian workers. *Labour & Industry, 12*(3), 43–58. http://dx.doi.org/10.1080/10301763.2002.10722023

Vezzoli, S. (2014). The effects of independence, state formation and migration policies on Guyanese migration. *International Migration Institute Working Paper Series, No 94.* Oxford.

4 "An Island Off the West Coast of Australia": Multiplex Geography and the Growth of Transnational Telemediated Service Work in Mauritius

CHRIS BENNER AND JAIRUS ROSSI

Introduction

> As I listen to phone calls here, people ask [the agents], "Where are you?" They know they're not talking to India ... At first ... the agents would say, "Mauritius." Well, most Americans don't know where Mauritius is ... and they say, "Where's that?" Some [agents] would say, "Oh well, it's off the coast of South Africa," but the more positive response that was very funny that they got is, "We're off the west coast of Australia." I've also heard Mauritians talk about that they're located just south of India. I'm like, "That's one you wanna avoid with Americans right now." (U.S. business client of a call centre in Mauritius, June 2007)

At first glance, the notion that Mauritian call centre agents can successfully convince typical American consumers that their country is near Australia seems to be another example of the limited geographic knowledge of the typical American consumer. Mauritius, after all, is a *long* way from Australia – some 3,500 miles – and has few cultural or political ties to this distant Indian Ocean "neighbor." In geographical and cultural proximity Mauritius is much closer to Africa and India or even its former colonial rulers, France and Britain. Yet the idea that call centre agents can claim to be simultaneously off the coasts of Africa, India, and Australia captures one of the central features that helps to explain the growth of outsourcing in Mauritius: the multiplex character of the island's cultural, economic, and political geographies.

Since the Internet revolution of the 1990s many developing countries have attempted to attract international call centres, business processing outsourcing (BPO), and other forms of transnational telemediated service work (Datamonitor, 2004; M. Graham & Mann, 2013; Kim & Nelson, 2000; Wayde & Rogerson, 2014; Wilson & Wong, 2006). Many countries in Africa see compelling potential for this rapidly growing, labour-intensive service work. By leveraging expanded telecommunications links, many hope to escape primary commodity dependence and to "leapfrog" stages of development into advanced services (Davison, Vogel, Harris, & Jones, 2000; Hyde-Clarke & Van Tonder, 2011; Oshikoya & Hussain, 1998; Semali, 2012). While most countries in Africa have struggled, since the establishment of its "cyber island" initiative in 2001 Mauritius has experienced double-digit growth in the numbers of firms and employees in its broad BPO sector, with evidence of increasing diversity, complexity, and dynamism of operations. How has this tiny island nation been able to grow these kinds of jobs?

Clearly, one of the important factors has been Mauritius's ability to meet certain prerequisites for competing successfully in information-technology-enabled service industries. It has a robust telecommunications infrastructure, a stable political and business climate, and an adequately skilled workforce willing to work for a cost that competitively matches the quality of the services they provide. But these qualities are hardly unique. What really sets Mauritius apart is the way in which industry and government leaders strategically inflect and shift the island's multiplex national identity to resonate with diverse international clients.

This chapter argues that an understanding of Mauritius's "muliplexity" of cultures, histories, political economic ties, and diasporic networks provides valuable theoretical insights into the dynamics of the country's transnational service growth. The conclusions drawn here are based on a total of thirty-two in-depth interviews with call centre managers and industry association executives conducted in June 2007, along with a detailed review of company data, industry reports, newspaper articles, and other secondary sources. Potential respondents were identified through snowball sampling and were chosen to reflect as wide a perspective on employment opportunities and conditions in the industry as possible. No one who was contacted declined to be interviewed. Interviews were semi-structured and open ended, using an interview protocol as a guide, and each lasted from a minimum of forty-five minutes to a maximum of three hours. We should emphasize that we did not interview call centre workers directly, and our analysis is limited to what we see as the factors shaping the growth of call centre jobs; it does not include the experience of workers themselves, though we do examine indicators of job quality such as average wages and opportunities for advancement.

We begin with a review of multiplexity, its previous applications in economic development, and its relevance in analysing transnational service work. In particular, we define the concept of *multiplexity* as encompassing (though not limited to) a place's multiple inflections of national identity. These identities continually mediate and are mediated by the transnational relationships that are made possible by advanced communication infrastructures and historical social relations. We then contextualize Mauritius's relative success in promoting these transnational service industries, within a general overview of other efforts in African countries. This is followed by an analysis of the way in which four different aspects of Mauritius's multiplex nation have contributed to the growth of transnational service opportunities: linking territorial extensibility and relationship proximity; building opportunity out of heterogeneity; developing innovation out of dense interactivity; and strengthening integration through a complex institutional structure that weaves together the state, the economy, and civil society. We conclude with some reflections on the experience in Mauritius, which, despite its unique characteristics, suggests that an appreciation of multiplexity in national identities, relationships, and diasporic networks could help in understanding the patterns of growth in transnational service work in other countries.

Multiplexity and Economic Development

The term *multiplex* at its most basic level simply means "having many parts." Its most common scientific usage, in the area of telecommunications and computer networks, refers to processes whereby multiple analogue message streams or digital data streams are combined into one signal over a shared medium. Distinct signals are combined and sent over a fibre-optic cable or electronic wire and later parsed into their component bits or electromagnetic waves with the constituent messages decoded. These processes allow for an efficient use of expensive communication infrastructure because they allow the simultaneous transmission of multiple messages or signals on the same circuit or channel (Thiele & Nebeling, 2007).

It is this notion of a single medium, with multiple parts that are sometimes distinguishable and sometimes not, providing connections between multiple sites of communication and interaction, that we want to particularly emphasize here. The medium can be material like a fibre-optic cable or more intangible like a personal relationship, an economic network, or a national identity. In this paper, Mauritius is the medium, transmitting multiple incarnations of itself to assert connections to, and to facilitate relations with, other places within a competitive global economy.

While notions of multiplexity have an older history in sociology and social network analysis,[1] within economic geography Amin and Graham (1997) provide one of the early specific uses of the term *multiplex*. In their paper ironically titled "The Ordinary City" they point out major limitations in the then-prominent approaches to understanding the revival of cities in the context of globalization, arguing that analysts miss important aspects of urban development by not embracing cities' multiplexity – instead, they over-generalize from paradigmatic examples or over-emphasize particular spaces or partial representations of the city. Amin and Graham argue that we need to understand cities as "the co-presence of multiple spaces, multiple times, and multiple webs of relations, tying local sites, subjects and fragments into globalizing networks of economic, social and cultural change" (pp. 417–8). They argue that a focus on multiplexity stresses the interconnections between the multiple time-space circuits and dimensions of urban life and helps to improve our understanding of the way in which the tensions and synergies between these differences help to shape urban dynamics (Amin & Graham, 1997). This approach, which was developed further in Graham and Marvin's path-breaking 2001 book *Splintering Urbanism*, argues that cities are best understood not as territorial units but as "a set of spaces where diverse ranges of relational webs coalesce, interconnect and fragment" (Amin & Graham, 1997, p. 418), or as "staging posts in the perpetual flux of infrastructurally mediated flow, movement and exchange .... constituted through many superimposed, contested and interconnecting infrastructural 'landscapes'" (S. Graham & Marvin, 2001, p. 8).

The focus on multiplexity is particularly important in understanding economic relations. Numerous writers after the "cultural turn" in economic geography have analysed the various ways that culture and identity shape economic dynamics (Barnes, 2001; Schoenberger, 1997; Storper, 1997). At the core of insights generated from this approach is the understanding that markets are not simply sites of economic exchange. They are also simultaneously sites and spaces of emotional development, value formation, exercise of political power, transmission of cultural norms, information sharing, knowledge creation, social net-work construction, identity formation and affirmation (or negation), and so on (Smith, 2005).

Understanding the multiplexity of such economic ties is especially important in the context of transnational service work for two main rea-sons. First, the technological space in which the work takes place both enables the labour practices and generates information about those labour practices – both automating and "infomating" labour activities (Zuboff, 1988) – a dimension of multiplexity that is not present in non-digitally mediated work processes. In essence, work that takes place in a telemediated environment is, at least potentially, also completely transparent to outside observers through detailed recording, moni-toring, analysis, and other methods of codifying digital interactions. Dynamics of growth and change in transnational service work are thus fundamentally related to the dynamic between performing the work and making efforts to control, analyse, automate, or innovate the work based on the infomating potentials. Improved information technology, for example, both enables more complex interactions in a telemediated environment, such as transcontinental telesurgery (Kumar & Mares-caux, 2007), and allows automation of more routine activities, through voice-recognition systems, for example. At the same time, information generated from telemediated labour processes can be used either to empower and improve service workers' activities, through expanding opportunities for informed self-reflexive analysis of the labour process, or to build a nearly ubiquitous, external, "Taylorized" panopticon of monitored control – a virtual "assembly line in the head" (Taylor & Bain, 1999). The way these diverse tensions play out in various sites of interaction largely shapes the quantity, quality, and variety of transna-tional service work opportunities.

Second, the ability of telemediated infrastructure to transcend dis-tance creates possibilities for interactivity across more diverse places and cultures (Castells, 1996; S. Graham & Marvin, 1996). Such intercon-nections create dramatic new possibilities for economic change but also raise new communicative complexities that are still only poorly under-stood (Bryson, Daniels, & Warf, 2006; Wilken, 2011). The early days of remote transnational service work development, for example, led to extreme predictions of the massive relocation of interactive service jobs, given the dramatic wage differences in different countries. Yet, over time many firms that initially rushed to relocate work overseas have reversed those decisions after recognizing the difficulties of maintaining quality interaction across multiple cultures in interactive service work. Understanding the multiplexity of these telemediated interactions, with their economic, technological, social, and cultural elements, is essential for understanding the dynamics of growth in transnational service work.

Incorporating multiplexity into an analysis of the evolution of the transnational service work processes in a particular place, such as Mauritius, is challenging, given the complexity involved. Here, we return to four dimensions of multiplexity from Amin and Graham (1997) that provide a useful way of parsing and comprehending the diverse world of transnational service work in Mauritius.

The first dimension is relational proximity and its interaction with time-space extensibility. Their central assertion is that territorial proximity does not necessarily equate to meaningful relationships, with communication technologies making the maintenance and deepening of meaningful relationships over a distance highly possible. Intense face-to-face interactions within urban space coexist with the dynamics of technologically mediated flows of communication and contact over increasingly distant networks. How do the complex interchanges between place-based relational webs and distantiated ones affect economic change? As Amin and Graham argue, "the nature of the balance is likely to be critical in assessing urban creativity, the propensity of cities to innovate and the complex interactions between economic 'competitiveness' and social 'cohesion'" (p. 419). In the context of transnational telemediated service work in Mauritius, the concept of multiplexity provides a useful way of understanding the diversity of the diasporic networks that intersect on the island and facilitate relational proximity to multiple other sites, while remaining rooted in a multicultural integrated territory.

Second is the relationship between heterogeneity and opportunity. Amin and Graham argue that "heterogeneities are essential to the dynamics of the contemporary 'urban'… [and] are, simultaneously, sources of economic dynamism and cultural innovation, and pointers to new notions of urban governance and institutional innovation" (p. 419). Analysing the relationship between heterogeneity and opportunity in transnational service work in Mauritius provides insights into both the different types of transnational service work processes that have developed on the island, and the way in which this diversity of labour processes builds multiple opportunities, for workers, firms, and the country.

Third is the relationship between place density and innovation. Innovation is essential for economic success in capitalism, and this dynamic is particularly strong in information-based industries where the processes of "ubiquitification" are accelerated (Maskell, Eskelinen, Hannibalsson, Malmberg, & Vatne, 1998; Maskell & Malmberg, 1999). Dense places tend to facilitate processes of innovation, particularly around assets, such as untraded interdependencies and common conceptual frameworks, that are difficult to imitate (Storper, 1997). There are signs of innovation in transnational service work activities in Mauritius, yet an examination of the relationship between density and innovation highlights the particular tensions and challenges facing further development.

Fourth, Amin and Graham argue that the institutional complexity of urban areas is also important, particularly as attention has shifted from hierarchical government processes to more complex webs of urban governance, which can either contribute to greater fragmentation and marginalization or build complex interdependencies among multiple urban actors. They assert the "crucial importance of decentred, integrative, and interactive governance styles within the broader dynamics of the multiplex city" (p. 420). The institutional structures and dynamics in Mauritius exemplify such decentred, yet integrative and interactive governance processes, which provide a coordinated flexibility in developing transnational service work in the country, while ensuring a broad-based and shared prosperity that has helped Mauritius to avoid largely the enclave nature that characterizes transnational service work in other locations.

Overall, we argue that looking at these four dimensions of multiplexity in Mauritius provides valuable insights into the country's ability to develop and expand transnational service work opportunities. Before moving into specific analysis, we first provide some background on trends in transnational service work globally, in Africa, and specifically in Mauritius.

Transnational Telemediated Service Work in Africa and Mauritius

The term *transnational telemediated service work* or *telemediated work* refers to all labour processes involving the transmission of voice and digitized information (including words, numbers, images, and computer programs) via a telecommunications link. This can include software development, data processing, sales, customer service, technical support, web design, and many human resource, training, creative-design, and research-and-development activities (Huws, 2003). While the particular business functions may vary, the actual work itself is shaped by the tel-emediated infrastructure, with work performed in one place linked to real-time interaction with customers located in another.

The locational dynamics of this telemediated work are now following the earlier relocation processes of manufacturing and non-interactive, back-office functions from the global north to the global south (S. Graham & Marvin, 2001). The factors shaping the location, quality, and character of telemediated work activities, however, are highly complex and still poorly understood (Massini & Miozzo, 2012). Some factors – such as labour costs, infrastructure, tax and regulatory structure – shape location and growth trajectories in ways that are similar to those of manufacturing industries (Stevens, 2014). Yet the interactive element of telemediated work brings in a wide range of cultural, linguistic, quality, communicative, and learning-process factors that potentially remain more important than cost factors in shaping location patterns (Liu, Feils, & Scholnick, 2011).

Given the challenges of measuring this type of work, global statis-tics on the industry are inconsistent, and most estimates of employ-ment in developing nations come from international consulting firms such as A.T. Kearny, Datamonitor, and McKinsey & Associates, which must be regarded with some scepticism, given their economic interest in exaggerating the size of the industry. Nonetheless, their estimates show that global offshoring of this type of work was growing in revenue by between 30 and 60 per cent a year in the mid-2000s. McKinsey & Asso-ciates estimated that global industry opportunities were on the order of $50 billion–$120 billion, with a total of two million people employed around the globe in the provision of telemediated service work services to clients based in another country (McKinsey Global Institute, 2003). In 2013, KPMG estimated that the global market was $950 billion, includ-ing information-technology (IT) professional services (which was $309 billion on its own), and the IT outsourcing market was $288 billion, up from $246 billion the previous year.[2] For employment, in 2003, McK-insey estimated that India was likely to have 1.1 million of those jobs, with China and the Philippines accounting for another 400,000 to 700,000 jobs. This left a gap of 200,000–500,000 jobs likely to be filled by other countries such as Malaysia, Singapore, Russia, Thailand, and two African nations – South Africa and Mauritius – that were identified as likely growth sites (McKinsey Global Institute, 2003). By 2014, India alone was estimated to have more than three million people employed in the global IT and BPO market (Manyika et al., 2014, p. 75). The McKinsey

Global Institute estimates that up to 160 million jobs could be performed remotely (McKinsey Global Institute, 2005).

Many countries around Africa have been trying to realize growth in transnational telemediated service work and are gaining increasing visibility. In 2014, A.T. Kearney included, in the fifty top countries in the world for offshoring of service activities, seven African nations (Egypt, tenth; Tunisia, twenty-eighth; Ghana, twenty-ninth; Morocco, thirty-fourth; Mauritius, thirty-sixth; Senegal, fortieth; and South Africa, forty-seventh) (A.T. Kearney, 2014). An article in *InfoWorld Magazine* prominently proclaimed in 2006: "Africa to lead call center growth" (Gohring, 2006). A number of African governments, including those of Ghana, Kenya, Botswana, South Africa, and Mauritius, have placed attracting transnational service work as one of three or four strategic industries for national economic policy.

However, outside of South Africa (Benner, 2006) and Mauritius, significant success to date in sub-Saharan Africa has been elusive. Ghana has one large multinational firm (ACS) that employs over two thousand people doing data-entry, a locally owned firm (Rising Data Solutions) that grew in fits and starts in call centre work before going out of business after the global recession, and a small number of software firms, but it has struggled with high telecommunications costs, poor building and power infrastructure, and surprisingly high labour costs for the quality of service. Kenya received a lot of publicity with the launching of KenCall, a call centre firm, but growth seems to be limited. Botswana launched a major initiative to attract financial services but has yet to attract or grow significant international business (Benner, 2007, 2014).

South Africa has had some success, building on a domestic base of financial services and call centre firms. Total employment within call centres and BPO operations in South Africa was estimated in 2009 to be around 100,000, and the government expected significant growth in international businesses following a new BPO growth initiative. However, only approximately 25%–30% of its total telemediated service workforce is estimated to be serving an international market, with the rest serving the domestic market. The impact of the government's incentive program has been significantly less than was originally envisioned (Benner, 2006; Benner, Lewis & Omar, 2007; Cape & Touche, 2008; Jones, 2009).

Mauritius's experience in the growth of transnational telemediated service work contrasts with the relatively slow pace of growth in other African countries. The first firm was established in Mauritius in 1996, and another six were established between 1998 and 2000. However, the government began to actively promote IT industries in 2001, launching a high-profile "cyber island" initiative. The program included efforts to expand broadband technologies, increase marketing, and provide specific incentives to firms engaging in international telemediated service work, and the creation of Business Parks of Mauritius Ltd. (BPML), with the goal of creating a "cyber city" to further attract transnational service work enterprises.[3] BPML quickly leased out its first twelve-storey, 42,000 square metre "cyber tower," with its cutting-edge technological infrastructure, and by 2007 it was busily filling a second, completed cyber tower.

The real take-off in Mauritius's transnational service work industry, as shown in figure 4.1, emerged following the opening of the SAFE (South Africa–Far East) fibre-optic cable in May 2002. This cable, which links South Africa to India and Malaysia via Mauritius, also connected to the SAT3/WASC (South Atlantic 3/West Africa Submarine Cable) that links South Africa to Europe via the west coast of Africa (Malcalm, 2006). It provided Mauritius with the broadband capabilities required for conducting reasonably priced voice and complex interactive communication from a distance. Prior to having this infrastructure in place, international transnational service work in Mauritius was destined to remain small and marginalized (Sokolov, 2005).

In 2003, growth started to take off, with another twenty-four firms being established in that year alone, doubling the number of firms and employment in the industry as a whole. By June 2004 the BPO Secretariat had identified sixty companies, who were employing a total of nearly 2,300 people in international telemediated service activities, and growth in both companies and employment continued at a dramatic rate. The number of firms operating in the country more than tripled in four years, while employment had grown by 167 per cent by 2007. By 2013 the Board of Investment Mauritius (BOI Mauritius) had counted a total of 631 transnational service companies operating in the country, employing over 19,000 people, contributing 6.4 per cent of the country's gross domestic product, and accounting for $170 million in export earnings (BOI Mauritius, 2014). There were also signs of increasing complexity and diversity of the industry over time, as companies moved up value chains to more complicated work.

Although the development of the multiplexing telecommunications infrastructure was an essential prerequisite for this rapid growth, in the following section we will argue that it is the multiplex geography of Mauritius that has been a major factor in contributing to the country's success to date.

Figure 4.1 Firms and Employment in the Transnational Telemediated Service Industry, Mauritius, 2004–13.

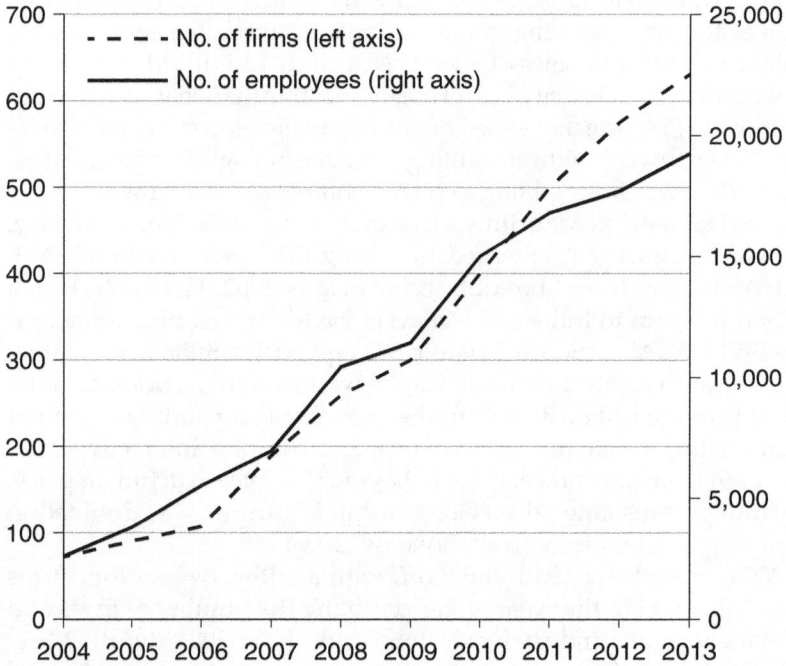

Source: BOI Mauritius (2014).

Multiplexity and the Growth of Transnational Telemediated Service Work in Mauritius

Mauritius's multiplex geography is linked in part to its physical location and in part to its unique history that placed it as a node in multiple diasporic networks. Situated in the Indian Ocean between South Africa and the southern tip of the Indian subcontinent, Mauritius is considered part of sub-Saharan Africa and has been a member of the African Union since independence in 1968. The island, though, has extensive links to Asia, with a majority of its population tracing its heritage to India. In fact, in recent years it has increasingly tried to market its Asian connections to both India and China, using such imagery as an [Asian] "Tiger in Paradise" or "Pearl of the Indian Ocean" (Aumeerally, 2005). At the same time, it maintains strong cultural links with its former colonial powers of England and France. Furthermore, none of the island's multiple ethnic groups can lay a claim to being indigenous; prior to colonization Mauritius was devoid of human life.

Although Arab traders apparently visited the island as early as the tenth century, when the Portuguese first landed on the island in 1510, there were no signs of human habitation. The Portuguese never established a lasting settlement on the island – the Dutch were the first true residents, establishing settlements from 1638. The Dutch left no lasting legacy on the island other than driving the dodo bird to extinction. France claimed the island in 1715 and began to shape its economy and culture. French colonists established large sugar-cane estates, using slave labour from the east coast of Africa and Madagascar (Meisenhelder, 1997).

The British took control of the island in 1810, though the French oligarchies were allowed to retain control of large portions of the sugar-production industry. After the British abolished slavery in 1835, African slave labour on the sugar plantations was replaced with indentured workers from South Asia and China. Significant numbers of free merchants, from northwestern India and China, also migrated to Mauritius during this period to provide goods and services to the indentured workers. This diaspora provided the origins of mercantile trading networks with India and contributed to the growth of an Indian middle class.

There are no longer precise statistics on the ethnic composition of Mauritius's population,[4] but it is estimated that Hindu Indo-Mauritians account for 52 per cent of the population, Creoles (mixed descendants of African slaves) 27 per cent, Muslim Indo-Mauritians 16 per cent, Sino-Mauritians 3 per cent (though 10 per cent of the population of the capital city, Port Louis), and Franco-Mauritians 2 per cent (Srebrnik, 2000). The country is officially bilingual (French and English), but 87 per cent of the population speaks Creole at home. The second-largest language spoken at home is Bhojpuri (sometimes considered a dialect of Hindi), with 5 per cent, and French is third at 4 per cent (Statistics Mauritius, 2011). It is not simply that the island is multi-ethnic but that the particular nature of the development of this multiplex geography has contributed to the four previously mentioned dimensions of multiplexity in transnational service work in Mauritius.

Relational Proximity, Distantiated Networks, and Multiple Diasporas

Critical to the success of transnational service work is relational proximity across distantiated networks at the levels of both the front-line worker and the executive. For front-line workers, the ability to establish productive telemediated communication with customers overseas is critical. The importance of this connection is most obvious in call centres, where the quality of the interaction depends on quite subtle cultural competencies for which it is difficult to train. Mauritians' ability to build these cross-cultural, transnational relations is facilitated by the extensive and multiple diasporic networks that have been built up over the many years of Mauritius's history (Srebrnik, 1999). Ethnic Chinese, South Asians, and French transcended the geographic isolation of Mauritius to tie the island to a global system of production. While the British settler population was always small, colonial relationships created extant diasporic connections with the United Kingdom.

These diasporic networks have been important in building transnational service work connections. On the simplest level, this has to do with the diversity of the ties. Most former colonies have inherited a single colonial legacy, with a single language. For former British colonies, this limits their market opportunities for transnational service work primarily to English-speaking countries. The dominant markets for India, South Africa, Philippines, and Ghana, for example, are the United States and the United Kingdom. For Senegal and Tunisia, France is almost their exclusive market. For many countries in Europe without that colonial legacy (for example, Scandinavia), there are very few opportunities for relocating transnational service work to lower-cost locations overseas.

Mauritius has the "good fortune" of having had two colonial powers and subsequently two official languages. Despite English colonial rule from 1810 until independence in 1968, French remains a dominant language and is fluently spoken more widely than is English. Nevertheless, all high-school graduates learn and display competence in both. Beyond linguistic abilities, the mix of French and English in the particular Mauritian multicultural context has led to Mauritian accents that are unique and frequently unfamiliar to remote customers. In the current context of the large-scale relocation of transnational service work to India (from the United Kingdom and the United States), and the somewhat smaller relocation to North Africa (from France), this accent facilitates connections, as reflected in the following statements by a call centre client and by a founder.

And in a positive way, [Mauritians] speak fantastic English comparatively in terms of an accent that an American can comprehend, and I think a big part of that is because even though it is British English with the French twinge, being in the States we have the Creole accents that come up from the Caribbean – Americans have kind of heard that. Their accents almost sound like a sweet Jamaican or Barbadian accent … They know they're not talking to India. (U.S. business client of an English-speaking call centre, June 2007)

They will recognize an absence of accent. They can't place it. It's not a French regional accent. It's not an accent they are used to, but fairly enough it is not an accent which they resent … for example, it is not a North African accent. (Founder of a large French-speaking call centre, June 2007)

At the executive level these transnational connections are also important. Establishing business opportunities in a new environment requires a level of trust, which is facilitated by personal connections, social networks, and cultural similarity. India's success in building telemediated work, for example, was built originally on social networks established by Indians working in the United States and leveraging these ties into ongoing business opportunities (Saxenian, 2006). Many transnational service companies that were started in Mauritius emerged through the personal experiences of living and working overseas that are rooted in these diasporic networks. The following are some typical examples of networks among informants in our sample.

- *Two (non-white) Mauritian-born co-founders of a large French-speaking call centre.* One studied in the United Kingdom and worked for Deloitte & Touche and then British American Tobacco, helping to install and maintain their IT systems (which included stints in Mauritius, Belgium, and Kenya). After many years he left that work to start his own company, distributing Apple computers, and later Hewlett-Packard equipment. The other founder was born in Mauritius but moved to France as a child. After studying in Paris and working for IBM, and later SSI (Société de Services Informatiques) as a software-development engineer, he returned to Mauritius and founded his own software-development firm in 1999, doing work for previous contacts in Paris. After growing his company, and selling it after three years, he joined his co-founder in starting a call centre – again building on customer contacts made through their previous time overseas.
- *Three founders of a BPO firm serving primarily a French market.* One founder was French and married to a Mauritian woman. He had worked in France for a company that was performing outsourced services within France for other companies. He met the other two founders during a "pre-retirement"' holiday in Mauritius.

- *A Swiss-owned financial services firm.* The firm was started through the personal networks linking a Mauritian living in Switzerland and the parent company.
- *A large document services firm.* This firm was established originally in the 1980s and was later bought out by a major British firm. It traces its origins to an entrepreneur in the United Kingdom who had links to the cargo division of a major international airline. He heard about Mauritius through the airline operations and knew someone from the island who was working for the airline. He asked that individual, who happened to know the Mauritian minister of tourism, to arrange an introduction to the minister when he was next in the UK. The entrepreneur then came to Mauritius and established the business over the next six months, again starting from customers he had known from work in the UK.

These stories emphasize the importance of social networks that include former colonial powers. Yet Mauritius also benefits from the diasporic networks of Indians, Chinese, and South Africans, who are able to provide needed capital and expertise. Although many South Asians were forced to migrate to Mauritius, free Indian merchants established extensive trade networks between the Indian subcontinent and Mauritius. These migrants focused on trade and the provision of services such as finance and moneylending for indentured and free Indian populations. As Indo-Mauritians gained land and influence on the island, the trading networks expanded to encompass the Indian subcontinent and other diasporic Indian communities such as the large enclave in Natal, South Africa (Markovits, 1999).

In Mauritius these historical linkages have lead to reciprocal investment between the island and India in the transnational services and IT sectors. Indian companies such as Infosys, Satyam, and Tata have invested in Mauritius and are viewing the island as a strategic location to build ties to other African countries. The Indian infrastructure giant Tata was involved in the construction of the underwater SAFE fibre-optic cable that connects Africa and Asia through Mauritius.

Chinese diasporic networks have also been important in establishing transnational business networks. Chinese merchants came to Mauritius as early as 1750 while involved in joint trading operations with the French. On the island, severing ties to one's homeland was a precondition of land ownership. Most Chinese residents refused this condition and thus were limited to activities related to trading. In 1900, 80 per cent of all Chinese were traders, and this has resulted in a continued relationship with Taiwan, Hong Kong, and mainland China (Brautigam, 2003). Chinese entrepreneurs were vital in attracting investment to the export processing zones (EPZs) in the 1970s, and this modern economic relationship continues to expand with China's rapid growth. Small to medium-sized enterprises (SMEs) in Mauritius often were joint partnerships with foreign investors located in ethnic Chinese lands. Since at least the 1990s Mauritius has positioned itself as a jumping-off point for investment in production in Southern Africa by Chinese, Taiwanese, and Hong Kong investors. The investors seek to ally with Sino-Mauritians to take advantage of trade agreements and economic coordination within the SADC (South African Development Community) and COMESA (Common Market for Eastern and Southern Africa) blocs (Brautigam, 2003; Srebrnik, 1999). While Chinese investment in transnational service work activities has been more limited, it has begun to expand. Huawei, a prominent Chinese telecommunications provider, first established operations in Mauritius in 2002, selling telecommunications equipment. As of 2008 it had started a transnational service operation providing financial shared services. There is also a Hong Kong–based company called Bardo Mauritius that provides international Internet payment processing.

In both the Indian and the Chinese cases, Mauritius is considered a connection point between Asia, Europe, and Africa. This small island nation has many attributes that are beneficial in establishing such links. Historical trading networks and diasporic relationships have allowed Mauritius to be relationally situated in three continents simultaneously. Strong connections to each continent are necessary for transforming the island into a financial and IT hub. Mauritian ethnic networks encompass two of the fastest-growing economies in the world (India and China) and have the potential as a finance centre to link them with other established global powers. With many Asian investors hoping to expand into Africa through Mauritius, the country's remote, yet connected location can also facilitate economic infiltration into the African continent.

Heterogeneity and Opportunity

Another element of the relative success of transnational service work activities in Mauritius is the diversity of operations established on the island, including call centres and back-office, business-process, outsourcing operations – and, increasingly, software development, multimedia work, and a range of higher value-added operations. This heterogeneity of

Figure 4.2 Segmentation of Transnational Telemediated Service Firms in
Mauritius, 2013.

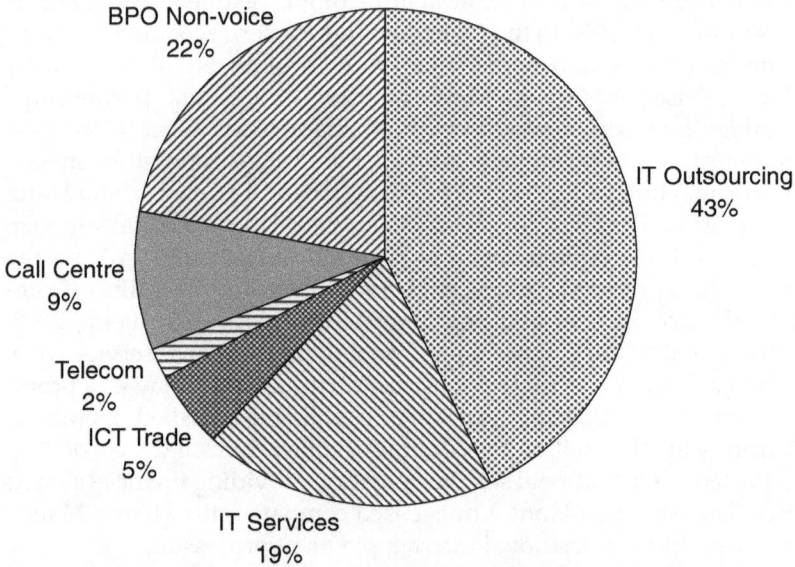

Source: BOI Mauritius (2014).

operations provides a variety of employment opportunities and at least
the possibilities of building cross-firm career ladders, which are important given the relatively limited advancement opportunities within call
centres and lower-end BPO operations (Gorjup, Valverde, & Ryan, 2008).
Over time the mix of operations in Mauritius has become more complex,
so that by 2013 some 67 per cent of transnational service firms in Mauritius were involved in the work of information and communications technology (ICT), while only 31 per cent were involved in the generally less
complex work of call centres or non-voice BPO operations (see figure 4.2).

An example of relatively simple transnational service work activities
is FrenchData, a division of a French firm that was established in Mauritius in 2003, which has entirely Mauritian management with substantial
international experience in South Africa and France. By June 2007 it had
grown to employ 150 people to do simple data entry. The parent firm
in France scans handwritten documents that need to be entered into
a database. Agents in Mauritius read the scanned image and type the
information into appropriate fields. Most of the work is in banking and

insurance, and a large portion of the documents are cheques. Although the company had plans to expand its operations to up to four hundred people, improvements in optical-character-recognition technology are likely to automate large portions of this work in subsequent years, making it a risky long-term business prospect. An average employee in the firm made Rs 6,000 (Mauritian rupees) a month in 2007, about the equivalent of $190 at exchange rates at the time.

An example of a transnational service work operation that provides more mid-skilled work is FrenchDTP, a page-layout and desktop publishing firm. As with FrenchData, the company was started as a subsidiary of a French firm in 1996 by a Mauritian founder, though by one who had been educated in England. The firm is linked to another subsidiary of the French firm that is located in Madagascar, where most of the lower-skilled work of basic data capture is done, paying the equivalent of $80 a month in 2007. In the Mauritius operations of French DTP the agents are primarily doing page layout and desktop publishing, which includes the preparation of owner's manuals and product booklets for large French corporations and multinational firms. Their items are produced in thirty different languages, and many are sent to Romania for proofreading, where the French company has a small subsidiary with thirty employees who speak twenty-three different languages – essential for ensuring quality control in the layout of the different language publications. Agents in Mauritius require knowledge of Adobe FrameMaker and InDesign software, and skill in a certain level of design, while a few of the employees also have programming skills. Agents are paid Rs 8,000(about $250) a month. Some of the better operators at the centre earn up to Rs 18,000 ($580) a month.

Most call centres in the country would also be classified as being in this mid-level range, with wages ranging from a low starting monthly salary of Rs 7,500 to as much as Rs 18,000 a month ($240–$575) for more experienced agents in the better-paid call centres providing complex services.

An example of a more complex operation is SwissFinance, a division of a Swiss firm that has fifty people in Mauritius conducting detailed research on certain business aspects of the largest publicly traded companies in the world. The research involves Web research on the intangible assets of these major firms, and entry of the relevant information into a detailed database that includes more than nine hundred specific metrics of a company's intangible assets. It typically takes an agent five to ten days per company to find and enter all the relevant data, and SwissFinance aims to eventually have more than thirty-five hundred companies in its database, with the information constantly kept up to date. The database is sold to major financial investment firms to assist them in their investment strategies. Agents in this firm typically have a college degree, with a few having a master's degree in business administration. Monthly salaries start at Rs 13,000 ($420), and by three years, agents can expect to be earning twice that amount.

The presence of such diverse transnational service work activities is important for at least three reasons. For workers in these centres, it provides potential opportunities for career advancement. One of the major problems of many call centres and low-end data operations is that they can have dead-end jobs, with high levels of turnover. Yet the skills obtained in one operation are often highly applicable in other operations, and the presence of sites with more complex operations provides potential for cross-firm career ladders. Second, the set of diverse activities means that the country is less vulnerable to downturns in any particular sub-component of the industry. For example, even if travel-related transnational service work operations face a downturn, the dynamics in financial services or software development may not. Thus, while transnational service work represents an effort to diversify Mauritius's economy overall, the diversity within transnational service work activities is also important. Finally, given the vulnerability of all transnational service work activities to the disruptive impact of technological change, their diversity makes it more likely that, as technological change enables the growth of new transnational service work possibilities, while causing the decline of others, Mauritius will have a presence in those growing sectors.

Place Density and Innovation

The long-term success of transnational service work activities in Mauritius will depend on the ability of Mauritian companies, mostly SMEs, to innovate. Innovation in Mauritian companies, rather than in foreign companies, is important because they are more likely to be rooted in ways that benefit the country and are less likely to move to lower-cost locations over time. Mauritius will never have the kinds of large, globally competitive companies that India has been able to develop. Yet building such large companies is not necessary for Mauritius to be successful. In global markets, specialized niches proliferate and create opportunities for smaller companies. There are signs that such innovation is occurring in Mauritius.

Consider, for example, the case of MUSmartCard, a Mauritian firm started by a Frenchman who had become a citizen after marrying a Mauritian woman. The company's main innovation lies in customizing smart-card[5] technology for an African market. Smart cards have certain advantages over the magnetic swipe technology used in most bank cards and credit cards. One advantage is security: magnetic cards are easily duplicated, and the information on smart cards can be more easily encrypted. A second advantage is the ease of reprogramming information onto the smart card. Taking advantage of these functions, MUSmartCard developed a payment service that targeted places in which telecommunications linkages were intermittent or non-existent, and therefore magnetic swipe cards were useless.[6] By contrast, money can be loaded onto a smart card at a central location, which can then be used to pay for services at any vendor with the appropriate card reader, even without a telecommunications connection. The transactions are recorded on the card and can be downloaded later to the back office in order to keep track of the transactions. The company has found applications in retail sectors – for example, with Kenyan Wildlife Services for payment of multiple services in wildlife parks – and in business services – for example, with gasoline companies delivering bulk fuel to service stations throughout the country. An additional advantage of this smart-card system is the ability to track electronic funds, rather than dealing with large volumes of cash. This is useful, for example, for tourists travelling in rural areas who know they cannot access their bank balances or pay with credit cards and do not want to carry large amounts of cash. It is also valuable to avoid "misplaced" cash, in both consumer and business interactions. By centralizing the processing through MUSmartCard servers, for instance, gas companies can avoid having their own employees deal with large quantities of cash when they deliver supplies to remote gas stations.

MUSmartCard has customers for its service throughout East and Southern Africa. It provides both the hardware (cards and terminals) and the service of processing the transactions. Roughly half of the company's Mauritian staff are high-level developers and programmers, making between Rs 15,000 and Rs 30,000 ($480–$970) a month, programming the cards and readers and customizing products for new clients. The other half are operators processing the financial transactions, who may start at a monthly salary of Rs 10,000 ($320) but can move to Rs 15,000 ($480); team leaders can earn up to Rs 25,000 ($800).

The company is small, with some thirty-five people employed in Mauritius and a handful spread around East Africa as field representatives.

But this small size, and the ability to find niche markets in underserved markets, provides a prototypical example of the sources of opportunities for innovation in Mauritius.

Transnational telemediated service operations in Mauritius are likely to remain dominated by SMEs, but this provides a competitive advantage in the niche market, compared to other countries that cater more to large multinational firms, as evidenced in the following statements by managers of call centres and business-process operations.

> Over the last five years it was large multinationals looking for huge providers globally, and they will go to India, they will go to Philippines, they will go to China and even South Africa, and say, "Okay, Mauritius is too small. We want 500 people, we want 1,000 people – in Mauritius we can't get that many." But after the big companies have been outsourcing, the medium-size company says, "Hey, they have done it, why don't we start doing it?" Now when the medium company goes out, they don't want to go to India, because when they meet a company in India and say, "I want 70 people," they say, "Oh, we are not interested," and they don't get the red carpet treatment, whereas when they come to Mauritius, we say, "Oh, welcome," because we want you … I have met a lot of people coming to Mauritius, and they say, "I have been to India; they are not very excited. When I come to Mauritius, I get a better treatment." (Manager of a British business-process outsourcing operation)

> So because of those requirements [the need to open quickly with a small operation] … I knew Mauritius was hungry for this type of business … If you come and say you employ 50, 60, or 100 staff in Mauritius in the call centre industry, you would have been met by the ministers! (Manager of a British call centre)

As shown in figure 4.3, Mauritius is the single largest country of origin for all telemediated services companies (mostly SMEs) operating in the country. SMEs in Mauritius have been specifically promoted through government policies since the 1980s, but can also trace their strength to the historical social organization of Indo-Mauritians (Lange, 2003). In 1913, following increased acquisitions of land by former indentured workers, credit unions and cooperatives were formed to protect the interests of the new small farmers. During the same period, villages organized into cohesive social units and acted to direct and distribute funds to social organizations and business enterprises.

Figure 4.3 Country of Origin of Investors in Transnational Telemediated Service Firms in Mauritius, 2013.

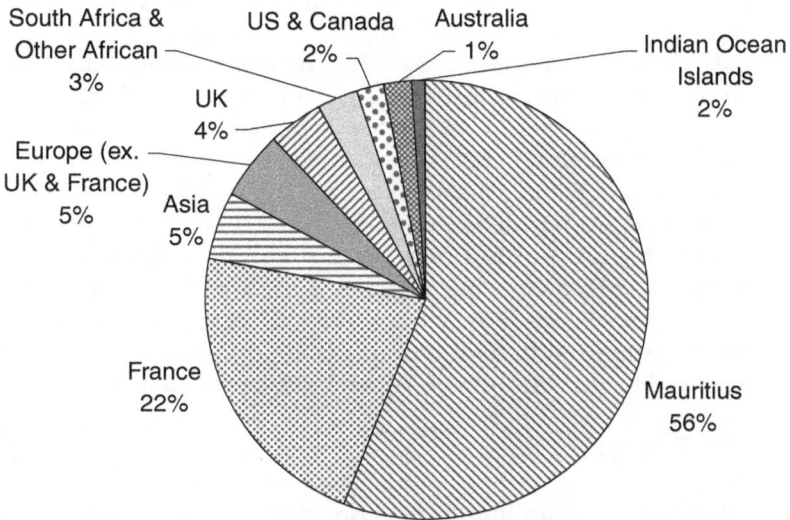

Source: BOI Mauritius (2014).

Small to medium-sized enterprises proliferated rapidly in the 1980s with the government promoting manufacturing enterprises in an effort to diversify the economy beyond its strict historical dependence on sugar. Activities in export processing zones (EPZs) expanded, along with related gains in the service sector, while tourism was embraced as part of this new economic trajectory. Mauritius declared the whole island to be an EPZ, providing preferential tax and regulatory benefits to new export companies. From its very origins the EPZ in Mauritius had strong domestic capital ties. In the early 1970s Sino-Mauritian entrepreneurs advocated the establishment of an EPZ on the island to mimic the success of newly industrializing countries in Asia. Travelling to East Asia, the entrepreneurs convinced private investors to invest in Mauritius's EPZs (primarily in textiles and other manufacturing). At the outset of the Mauritian EPZ, 90 per cent of the invested capital came from Hong Kong. As the zone slowly gained momentum, others invested (from China, Taiwan, France, and Mauritius), and by 1983 (when the EPZ operations really took off) Hong Kong's investment had been reduced to 59 per cent. Government SME-promotion acts supported investments in manufacturing in the form of genuine partnerships with Mauritian entrepreneurs. Thus, by the end of the 1980s, over 50 per cent of the equity in the EPZ was local (Brautigam, 2003). While foreign investment was necessary to stimulate this economic diversification, domestic capital investment led to the equitable redistribution of benefits to support the development of social and human capital.

The long-term promotion of SMEs has kept significant amounts of capital circulating within the island itself, which now provide a domestic base of investment for transnational service work operations (Larose, 2003). Mauritian innovation in both specific industry and macroeconomic state trajectories has come from a plurality of sources: multiple diasporic networks providing finances while reinforcing the investments with culturally based transnational relationships; diverse social networks overlapping with business and political organization on the island; and, as will be examined further, institutional complexity in decision making.

Institutional Complexity

The fourth dimension of multiplexity is the institutional governance of society. Here, unique combinations of cultural pluralism and national hybrid identity, decentralized flexibility, and coordinated actions, and the emergence of a networked developmental state are critical to facilitating economic development and ensuring a significant level of shared prosperity (Meisenhelder, 1997; Ó Riain, 2004). Since independence the state has played a leading role in promoting development. It used resources from the sugar industry to support social development and then implemented strategic initiatives at critical points in the country's history to help diversify the economy, first to support manufacturing enterprises in export-processing industries and later to promote tourism, financial services, and now transnational service work activities.

Since its independence in 1968 Mauritius has had an uninterrupted period of a stable, multiparty parliamentary democracy. Its multiparty system has been described as constituted by "a bewildering kaleidoscope of parties, often constituted for little more than electoral purposes" (Srebrnik, 2002, p. 277). Yet the class structure of the society, and its representation in state structures, has helped to promote domestic investments that ensure development opportunities. At independence the country's main economic engine, the sugar industry, continued to be the domain of French oligarchs, yet Indo-Mauritians also owned land and had smaller sugar estates. The origins of this middle class of

Indo-Mauritians can be traced to the 1870s, when cyclones, plant diseases, and dropping sugar prices on the world market catalysed the subdivision of large estates into smaller worker-run plots. These plots were predominantly purchased by indentured servants from India. This middle class of Indo-Mauritians figured prominently in the gaining of independence and achieved great influence in the post-independence bureaucracy.

After voting for independence in 1968 the island still maintained economic links with both the French and the British. The presence of a large French sugar oligarchy facilitated the negotiation of strategic trade agreements with European countries to fix sugar prices. The Lome Convention in 1975 established yearly fixed prices that were higher and more profitable than those of other sugar-producing countries (Brauti-gam, 1999). Instead of promoting a takeover of large francophone-run estates, the Hindu-lead government created a sugar tax that was levied only on estates of a certain size. The high revenues from this source were used to create broad social welfare programs including free education, employment, improved health care, and food subsidies during the 1960s and the 1970s, contributing significantly to social cohesion.

Mauritius has a strong tradition of social dialogue, bringing together business, government, and unions to negotiate economic policy. The trade union movement has played a significant historical role, in a sometimes bewildering array of individual unions and federations. As of the early 2000s there were some 330 unions in the country, grouped into eight federations, which were further organized into four main union confederations.[7] Despite this seeming fragmentation, all unions are represented in tripartite negotiations on labour, economic, and social issues, along with the Mauritian Employers Federation and the Mauritian government. One of the important institutions to emerge from this tripartite structure is the National Remuneration Board, which is responsible for prescribing the general conditions of employment, for fixing minimum wage rates, and for setting more specific remuneration rates in certain categories of employment. As of the early 2000s, the board had prescribed conditions of work for twenty-nine different sectors. Thus, while a minority of private sector workers are actually members of unions, the entire population is covered by annually negotiated increases in minimum wages, and the vast majority are covered by collective bargaining. Inequality certainly exists in Mauritius, and the Creole population in recent years in particular has been highly visible as having been left out of Mauritius's economic prosperity. Yet, with a

Human Development Index of 0.777 (in 2015) and a Gini index of 0.358 (in 2012), Mauritius ranks as one of the most equal and broadly prosperous countries in the global south (UNDP, 2015).[8]

Within the transnational services sector there is no single structure that provides overall governance. The central government has played a leading role in developing an overall strategy for transnational service work, yet implementation has occurred through more decentralized structures. The Board of Investment Mauritius has played a prominent role in marketing the country externally and continues to track development of the industry, including maintenance of an ongoing database of transnational services companies operating in the country, which is available on its website.[9] Business Parks Mauritius Ltd. was set up in 2001 and is owned by the government, but operates as a separate entity in building appropriate infrastructure and office space. The Human Resource Development Council was created by the government in November 2003 as part of the broad strategy to diversify the Mauritian economy. Though focused on a variety of different industries, it promotes skills training in BPO and other transnational service enterprises as one of its priorities.

In the private sector there have been efforts to develop an association of companies in the outsourcing and ICT sectors that have been somewhat less effective. Although the Association for Communication Technology was founded in 2004, its efforts were modest in the first three years. Members of the managing committee of the organization talked about goals of using the association to help support the growth of the industry as a whole, but they struggled to communicate the key accomplishments of the association. The organization was relaunched in September 2007 with a new name, the Outsourcing and Telecommunications Association of Mauritius, with somewhat higher visibility and involvement in various IT-related events, but with only twenty-two members and limited activities it remains far from being a vibrant source of support for the industry.

Conclusion

Although the growth of transnational service work in Mauritius has been remarkable since 2002, there are a number of factors that might limit the island's ability to promote new growth in this sector in the coming years. The factor most frequently cited in interviews was the constraints of the available workforce, in terms of both the absolute size of the workforce and the appropriate skills.

With a total workforce of some 548,900 in 2007, of whom 46,800 were unemployed, and only some 8,000 people employed in international transnational service work activities, it would seem that there is substantial room to expand employment in transnational service work activities. Those numbers, however, are somewhat misleading. Many of the unemployed are people who have been laid off from enterprises in the sugar industry, or in EPZ manufacturing enterprises, which have faced significant downturns in recent years. Most of these people lack the computer and language skills that are required for transnational service work activities. The primary source of new employees for transnational service work enterprises is recent secondary-school graduates or, for the more advanced positions, university graduates. In 2006, in the country as a whole, there were only 6,376 school leavers who passed the Higher School Certificate exam (out of 8,000 who took the exam), which is seen as a prerequisite for working in the transnational services industry.[10] As only a portion of these will be interested in pursuing transnational service work opportunities, constraints in absolute labour availability for transnational service work enterprises might be a barrier in years to come. Although none of the enterprises expressed current difficulties in recruiting people for entry-level positions, quite a few said that limited labour availability had already restrained the development of large (having more than one thousand employees) transnational service work enterprises, and they could see potential difficulties in the coming years if the industry continued to grow significantly.

The lack of skills at higher levels, particular in IT-related industries, may also become a constraint. Of the nearly 29,000 students in the country enrolled in higher education in 2005, only 4,100 were focusing on IT, and several of the higher-skilled enterprises in our interviews expressed difficulties in finding employees with adequate skills in computer programming, systems administration, and Internet design.

Mauritius may also face new competition in the future, from lower-cost sites that have similar linguistic abilities but lower labour costs. Madagascar is already seen as a potential new competitor. With a total population of nearly twenty million, it is more than ten times larger than Mauritius but has similar potential to serve a French-speaking market. There is also no guarantee that Mauritius will be able to continue on a trajectory of moving to higher value-added positions in transnational services. Many other countries (including those studied in other chapters of this volume) have expressed the hope of moving up a value chain from lower-paying to higher-paying transnational services, but with uneven success at best. The industry remains highly competitive, with continued rapid technological change and economic restructuring that provide few opportunities for the establishment of stable, higher-value, competitive niches.

Despite these real and feared threats Mauritius has significant potential for the continuing growth of transnational service work enterprises for years to come. The more positive dynamics in the industry can be linked to the multiplex character of Mauritius for a number of reasons. First, Mauritius's diverse cultures and diasporic networks open up a wide range of possible markets, and sources of investment, as they help to facilitate relational proximity despite Mauritius's remote location. Second, there is a wide diversity of transnational service work enterprises in Mauritius – in terms of countries served, types of operations, and complexity of the labour process – which is quite remarkable for the relatively small size of the industry. This also contributes to the potential for future growth. Third, Mauritian-led innovation in transnational service work activities potentially raises opportunities for moving to higher value-added activities. While the sources of this innovation are complex, the potential for growth likely lies in capitalizing on expanding markets in Africa and India, while continuing to leverage experience and resources from serving customers in Europe and the United States. Finally, Mauritius's complex governance processes can be effectively characterized as flexibly developmental and have helped to ensure that economic growth on the island has pursued a developmental trajectory, rather than simply extracting resources.

Are there lessons from Mauritius for other countries in the global south that are also trying to promote transnational service work? There are obviously features of Mauritius's multiplex geography that are unique, including the country's cultural heterogeneity, bilingual population, diverse diasporic networks, small population concentrated on a single remote island, and the particular advantages of a flexible developmental state, and which cannot be easily replicated in other developing countries. Nonetheless, there are some important aspects of Mauritius's experience that hold promise for other places as well.

One is the recognition that, despite India's and the Philippines' overall global dominance in transnational service work activities, there are opportunities for small countries to promote dynamic growth in transnational service work activities. A second point is that many developing countries also have valuable diasporic links to Europe and the United

States, which can become an important base for business opportunities, skills transfer, and investment finance. Finding better ways of supporting such diasporic networks and of helping to direct investment into transnational service work industry strategies might prove productive.

Finally, Mauritius's experience suggests the economic value of a flexible and multiplex conception of nation and nationhood. Many authors recognize the important distinction between state and nation, where *state* refers to a territorial unit in which the resident population is governed by an internationally recognized authority structure, and *nation* is recognized as a reasonably large group of people who see each other as sharing a common culture, history, and set of institutions. There are clearly many nations that are not coterminous with nation states, and there are many prominent cases in which internal diversity of national identity has led to political fragmentation (for example, Sudan and Yugoslavia, where internal national divisions have led to the creation of new states; or numerous countries in Africa and the Middle East, where internal divisions have made state governance challenging, to say the least). Yet what distinguishes Mauritius is not that the state has a diversity of national identities within its boundary but that the nation state's very national identity is rooted in a multiplex culture and transnational history in which diasporas beyond state boundaries are on some level seen as integral components of the nation state's identity. While Mauritius may be extreme in the muliplexity of its national identity, it is certainly not unique in having a history rooted in transnational migrations and the blending of multiple cultures. In an increasingly transnational and multi-ethnic world, learning how to turn such multiplex identities into economic assets, rather than into sources of political tension or fragmentation, seems to be perhaps the most important lesson Mauritius has to offer.

NOTES

1 Going back at least to the Chicago School, with its emphasis on multiple cultures and spaces of the city (Park, Burgess, & McKenzie, 1925; Wirth, 1928). Within social network analysis, see Burt (1983), McCarty (1996), Mitchell (1969), and Safford (2009).

2 http://www.statista.com/statistics/298574/bpo-and-it-services-market-breakdown-worldwide/ and http://www.statista.com/statistics/298554/global-spending-it-outsourcing-market/ (accessed 17 December 2014).

3 http://www.e-cybercity.mu/.
4 The census stopped gathering ethnic data in 1982 (Christopher, 1992).
5 Smart cards contain a microprocessor embedded in the plastic; they have become widespread in much of the world, though they are still less common in the United States.
6 A magnetic swipe card requires telemediated access to a centralized server with information about the cardholder's account.
7 The Mauritius Labour Congress, the Mauritius Trade Union Congress, the National Trade Union Confederation, and the Confédération Mauricienne des Travailleurs are those main union confederations.
8 Gini coefficient comes from the World Bank World Development Indicators. http://data.worldbank.org/indicator/SI.POV.GINI (accessed 9 April 2016).
9 http://www.boimauritius.com/.
10 Mauritius Central Statistical Office, http://www.gov.mu/portal/goc/cso/ei660/toc.htm.

REFERENCES

Amin, A., & Graham, S. (1997). The ordinary city. *Transactions of the Institute of British Geographers, 22*(4), 411–29. http://dx.doi.org/10.1111/j.0020-2754.1997.00411.x

A.T. Kearney. (2014). *A wealth of choices: From anywhere on earth to no location at all* (2014 A.T. Kearney Global Services Location Index). A.T. Kearney. Retrieved from https://www.atkearney.com/documents/10192/5082922/A+Wealth+of+Choices.pdf/61c80111-41b2-4411-ad1e-db4a3d6d5f0d (accessed 8 April 2016).

Aumeerally, N.L. (2005). "Tiger in paradise": Reading global Mauritius in shifting time and space. *Journal of African Cultural Studies, 17*(2), 161–80. http://dx.doi.org/10.1080/13696850500448261

Barnes, T. (2001). Retheorizing economic geography: From the quantitative revolution to the "cultural turn." *Annals of the Association of American Geographers, 91*(3), 546–65. http://dx.doi.org/10.1111/0004-5608.00258

Benner, C. (2006). South Africa on-call: Information technology and labour market restructuring in South African call centres. *Regional Studies, 40*(9), 1025–40. http://dx.doi.org/10.1080/00343400600928293

Benner, C. (2007, May). *International outsourcing and new forms of ework: An economic boom for Africa?* Paper presented at the Second International Economic Geography Conference, Beijing, China.

Benner, C. (2014). Digital dynamics: New technologies and work transformations in African cities. In S. Oldfield & S. Parnell (Eds.), *Routledge handbook on cities of the global south* (pp. 236–51). London: Routledge.

Benner, C., Lewis, C., & Omar, R. (2007). *The South African call centre industry: A study of strategy, human resource practices and performance.* Johannesburg: LINK Centre, University of the Witwatersrand.

Board of Investment Mauritius. (2014, February). Board of Investment eNewsletter. Port Louis, Mauritius.

Brautigam, D. (1999). Mauritius: Rethinking the miracle. *Current History (New York, N.Y.), 98,* 228–31.

Brautigam, D. (2003). Close encounters: Chinese business networks as industrial analysts in Sub-Saharan Africa. *African Affairs, 102*(408), 447–67. http://dx.doi.org/10.1093/oxfordjournals.afraf.a138824

Bryson, J.R., Daniels, P.W., & Warf, B. (2006). Service worlds: People, organisations, technologies. *Journal of Economic Geography, 6*(2), 246–7. http://dx.doi.org/10.1093/jeg/lbi007

Burt, R.S. (1983). Distinguishing relational contents. In R.S. Burt & M. Minor (Eds.), *Applied network analysis* (pp. 35–74). Beverly Hills, CA: Sage Publications.

Cape, C.T., & Touche, D. (2008). *Contact centres and business process outsourcing in Cape Town: 2007/8 Key indicator report.* Cape Town: Calling the Cape.

Castells, M. (1996). *The rise of the network society.* Cambridge, MA.: Blackwell Publishers.

Christopher, A.J. (1992) "Ethnicity, community and the census in Mauritius, 1830-1990" *The Geographical Journal, 158*(1), 57-64.

Datamonitor. (2004). *Global offshore call center outsourcing: Who will be the next India.* New York, NY: Datamonitor.

Davison, R., Vogel, D., Harris, R., & Jones, N. (2000). Technology leapfrogging in developing countries: An inevitable luxury. *Electronic Journal on Information Systems in Development Countries, 1*(5), 1–10.

Gohring, N. (2006). *Africa to lead call center growth.* Retrieved from http://www.infoworld.com/article/2655826/techology-business/africa-to-lead-call-center-growth.html

Gorjup, M.T., Valverde, M., & Ryan, G. (2008). Promotion in call centres: Opportunities and determinants. *Journal of European Industrial Training, 32*(1), 45–62. http://dx.doi.org/10.1108/03090590810846566

Graham, M., & Mann, L. (2013). Imagining a silicon savannah? Technological and conceptual connectivity in Kenya's BPO and software development sectors. *Electronic Journal of Information Systems in Developing Countries, vol. 56.* Retrieved from http://144.214.55.140/Ojs2/index.php/ejisdc/article/view/1107

Graham, S., & Marvin, S. (1996). *Telecommunications and the city : Electronic spaces, urban places*. London: Routledge. http://dx.doi.org/10.4324/9780203430453

Graham, S., & Marvin, S. (2001). *Splintering urbanism: Networked infrastructures, technological mobilities and the urban condition*. London: Routledge. http://dx.doi.org/10.4324/9780203452202

Huws, U. (2003). *When work takes flight: Research results from the EMERGENCE project*. Brighton, UK: Institute for Employment Studies.

Hyde-Clarke, N., & Van Tonder, T. (2011). Revisiting the "leapfrog" debate in light of current trends of mobile phone Internet usage in the Greater Johannesburg area, South Africa. *Journal of African Media Studies, 3*(2), 263–76. http://dx.doi.org/10.1386/jams.3.2.263_1

Jones, R. (2009). *National BPO & contact centre report*. Johannesburg: Contact Industry Hub. Retrieved from http://callcentrehub.com/researchfooter/47-the-national-bpo-call-centre-report

Kim, L., & Nelson, R.R. (2000). *Technology, learning and innovation : Experiences of newly industrializing economies*. Cambridge, UK: Cambridge University Press.

Kumar, S., & Marescaux, J. (2007). *Telesurgery*. New York, NY: Springer.

Lange, M. (2003) Embedding the Colonial State. *Social Science History. 27* (3), 392-423.

Larose, P. (2003). The impact of global financial integration on Mauritius and Seychelles. *Bank of Valletta Review, 28*, 33–49.

Liu, R., Feils, D.J., & Scholnick, B. (2011). Why are different services outsourced to different countries? *Journal of International Business Studies, 42*(4), 558–71. http://dx.doi.org/10.1057/jibs.2010.61

Malcalm, E. (2006). *Flattening the world: The prospect for fiber optic technology in Africa*. Iconnect-Online. Retrieved from http://www.witfor.org/2007/www.witfor2007.org/commission/building-the-infrastructure/Flattening_Ebenezer.pdf.

Manyika, J., Bughin, J., Lund, S., Nottebohm, O., Poulter, D., Jauch, S., & Ramaswamy, S. (2014). *Global flows in a digital age: How trade, finance, people, and data connect the world economy*. San Francisco, CA: McKinsey Global Institue.

Markovits, C. (1999). Indian merchant networks outside India in the nineteenth and twentieth centuries: A preliminary survey. *Modern Asian Studies, 33*(4), 883–911. http://dx.doi.org/10.1017/S0026749X99003467

Maskell, P., Eskelinen, H., Hannibalsson, I., Malmberg, A., & Vatne, E. (1998). *Competitiveness, localised learning and regional development : Specialisation and prosperity in small open economies*. New York: Routledge. http://dx.doi.org/10.4324/9780203276921

Maskell, P., & Malmberg, A. (1999). The competitiveness of firms and regions: Ubiquitification and the importance of localized learning. *European Urban and Regional Studies, 6*(1), 9–25. http://dx.doi.org/10.1177/096977649900600102

Massini, S., & Miozzo, M. (2012). Outsourcing and offshoring of business services: Challenges to theory, management and geography of innovation. *Regional Studies, 46*(9), 1219–42. http://dx.doi.org/10.1080/00343404.2010.509128

McCarty, C. (1996). The meaning of knowing as a network tie. *Connections, 18,* 20–31.

McKinsey Global Institute. (2003). *Offshoring: Is it a win-win game?* San Francisco, CA: McKinsey Global Institute.

McKinsey Global Institute (2005). *The emerging global labor market*. Part 1, *The demand for offshore talent in services*. San Francisco, CA: McKinsey Global Institue. Retrieved from http://www.mckinsey.com/global-themes/employment-and-growth/the-emerging-global-labor-market-demand-for-offshore-talent

Meisenhelder, T. (1997). The developmental state in Mauritius. *Journal of Modern African Studies, 35*(2), 279–97. http://dx.doi.org/10.1017/S0022278X97002383

Mitchell, J.C. (1969). The concept and use of social networks. In J.C. Mitchell (Ed.), *Social networks in urban conditions* (pp. 1–50). Manchester, UK: Manchester University Press.

Ó Riain, S. (2004). *The politics of high-tech growth developmental network states in the global economy*. New York, NY: Cambridge University Press. http://dx.doi.org/10.1017/CBO9780511499609

Oshikoya, T.W., & Hussain, M.H. (1998). Information technology and the challenge of economic development in Africa. *African Development Review, 10*(1), 100–13. http://dx.doi.org/10.1111/j.1467-8268.1998.tb00099.x

Park, R., Burgess, E., & McKenzie, R. (1925). *The city: Suggestions for the study of human nature in the urban environment*. Chicago, IL: University of Chicago Press.

Safford, S. (2009). *Why the garden club couldn't save Youngstown: The transformation of the Rust Belt*. Cambridge, MA: Harvard University Press.

Saxenian, A. (2006). *The new argonauts: Regional advantage in a global economy*. Cambridge, MA: Harvard University Press.

Schoenberger, E. (1997). *The cultural crisis of the firm*. Cambridge, MA: Blackwell.

Semali, L. (2012). Quest for economic empowerment of rural women entrepreneurs in Tanzania: ICTS leapfrom the digital divide. In R.N. Lekoko & L.M. Semali (Eds.), *Cases on developing countries and ICT integration: Rural community development* (pp. 91–103). Hershey, PA: IGI Global. http://dx.doi.org/10.4018/978-1-60960-117-1.ch010

Smith, S.J. (2005). States, markets and an ethic of care. *Political Geography, 24*(1), 1–20. http://dx.doi.org/10.1016/j.polgeo.2004.10.006

Sokolov, D.A.J. (2005). *Mauritius special report: A status check on progress to the "cyber island" dream*. London, UK: Balancing Act.

Srebrnik, H. (1999). Ethnicity and the development of a "middleman" economy on Mauritius: The diaspora factor. *Round Table, 350*, 297–311. http://dx.doi.org/10.1080/003585399108162

Srebrnik, H. (2000). Can an ethnically based civil society succeed? The Case of Mauritius. *Journal of Contemporary African Studies, 18*(1), 7–20. http://dx.doi.org/10.1080/025890000111940

Srebrnik, H. (2002). "Full of sound and fury": Three decades of parliamentary politics in Mauritius. *Journal of Southern African Studies, 28*(2), 277–89. http://dx.doi.org/10.1080/03057070220140702

Statistics Mauritius (2011). *2011 Housing and Population Census, Demographic and Fertility Characteristics, Excel Tables*. Mauritius: Statistics Mauritius. Retrieved from http://statsmauritius.gov.mu/English/Pages/2011-Housing-and-Populations-.aspx

Stevens, A.J.R. (2014). *Call centers and the global division of labor: A political economy of post-industrial employment and union organizing*. Abingdon, UK: Routledge.

Storper, M. (1997). *The regional world: Territorial development in a global economy*. New York, NY: Guilford Press.

Taylor, P., & Bain, P. (1999). An assembly line in the head: Work and employee relations in the call centre. *Industrial Relations Journal, 30*(2), 101–17. http://dx.doi.org/10.1111/1468-2338.00113

Thiele, H.-J., & Nebeling, M. (2007). *Coarse wavelength division multiplexing: Technologies and applications*. Boca Raton, FL: CRC Press. http://dx.doi.org/10.1201/9781420018691

UNDP. (2015). *Human development report 2015: Work for human development*. New York: United Nations Development Programme.

Wayde, R.P., & Rogerson, C.M. (2014). South Africa's call centre industry: The emerging challenges of a growing destination in the global south. *Mediterranean Journal of Social Sciences, 5*(8), 208.

Wilken, R. (2011). *Teletechnologies, place and community*. London, UK: Routledge.

Wilson, E., & Wong, K. (2006). *Negotiating the net in Africa: The politics of Internet diffusion*. New York, NY: Lynne Rienner.

Wirth, L. (1928). *The ghetto*. Chicago, IL: University of Chicago Press.

Zuboff, S. (1988). *In the age of the smart machine: The future of work and power*. New York, NY: Basic Books.

PART TWO

Constructing Nationally Appropriate (and Inappropriate) Workers

5 Serving the World, Serving the Nation: Everyday Nationalism and English in Philippine Offshore Call Centres

AILEEN O. SALONGA

Introduction

Convergys, one of the largest call centres in the Philippines, recently ran a local television commercial called "Pedestal" (Convergys, 2012). It opens with a scene in an urban centre. Many people are shown walking in different directions, rushing, and looking busy. As the scene fades, a horizontal structure made of concrete appears. The structure looks like a pedestal, but the top of it is empty. The scene then moves to what looks like a park. There are trees, and a young man is leaning on a concrete structure. As the man moves away from the structure, the camera moves up to show that it is also a pedestal, but instead of an empty top it has steel wires and dried leaves, suggesting that the structure on top has been uprooted for some time and perhaps forgotten. The scene then moves to a young couple, a man and a woman, taking a selfie. Behind them is another pedestal. The top is also empty. The scene once again moves, this time to show a man in a moving taxi. As the taxi moves, the camera also starts moving to show another empty pedestal within the vicinity. Then, finally, the scene moves to a group of young women who are eating, talking, and laughing in an alfresco restaurant in what looks like a seaside area. Again, the camera pans to an empty pedestal behind the women and the other guests in the restaurant. Meanwhile a male voice-over utters the following: "There is not one person in the country today who can say he's made a difference, not one who has taken that leap of faith to change his fate, not one who has made a sacrifice for his loved ones, not one who pushes this country forward. There is not one person today we can call a hero, not one."

Looking at the images and listening to the voice-over, the viewer begins to understand what the "Pedestal" commercial is trying to convey. All the pedestals are empty because there are no more heroes. And it looks like nobody really cares. The people in the commercial are shown to be busy with their own lives as they go about the daily grind, take pictures with loved ones, and eat and have fun with friends. Being heroes – or looking for them – does not seem to be a priority.

The "Pedestal" commercial, however, does not stop there. Something happens. Music is cued, and the male voice-over says: "But thousands." As this is said, another young man steps up and then crouches on a pedestal. He then begins to stand up with a determined look on his face and finally, standing, he looks up to the sky. The camera then starts showing many other pedestals in the background, but they are no longer empty. The commercial goes back to all the previous scenes, and in each one there are more pedestals and there are men and women standing proud and tall on each one. The people in the earlier part of the video – the men and women moving about, the man leaning on the pedestal, the couple taking a selfie, the man in the taxi, and the women in the restaurant – and a multitude of others are shown looking up at these people with great admiration.

In the end what the commercial wants to convey is that there is not one hero because there are thousands. The commercial is obviously playing on the idea that everyone can be a hero. This is, however, a television commercial. These so-called heroes cannot be anyone. As the men and women on the pedestals are shown more clearly, the camera zooms in on a wrist accessory that they are all wearing – a leather bracelet with a letter C pendant. In the background tall modern structures with the name Convergys are also shown.[1] At the very end of the commercial the name of the company is voiced over, "Convergys," and then the line "You count." Convergys is one of the biggest call centre companies in the Philippines; it was also one of the earliest to establish a centre in the country (Convergys, 2013b; O'Malley, 2013). Ultimately the commercial is selling the idea of call centre work as something positive, something to be desired, something that is, in fact, heroic. It is also selling Convergys as the ideal call centre workplace, the call centre that can help one be a hero.

There are many interesting aspects of the "Pedestal" commercial. Perhaps the most interesting, and also the most controversial, is the depiction of call centre workers as heroes. This is a bold move in light of the negative associations surrounding call centre work (for example, as not a real job because it mainly involves answering the telephones, or as a stressful job where one is abused by callers or customers) and call centre workers (for example, as young people who have a lot of purchasing power but are not really critical or discerning; as pretentious young people who use English with a fake American accent; or as victims of exploitative labour practices). However, this depiction of call centre workers as heroes is not really new. In construing call centre work as a heroic act, the commercial definitely counters the negative associations listed above, but, more crucially, it draws and builds on the *bagong bay-ani* (new or modern-day hero) narrative of overseas Filipino workers (OFWs), to whom the title *bagong bayani* was first attributed. The OFW-as-hero narrative is a government-authored narrative meant to recognize the sacrifices that OFWs make to be able to support their families, which in turn leads to the foreign remittances that help to keep afloat the Philippine economy (Cheng, 2010; Guevarra, 2010; Rafael, 1997; Rodriguez, 2002; Rupert & Solomon, 2006). This sacrifice for family and country makes the OFWs the modern heroes of the nation. There is no such government-authored narrative for Filipino call centre workers, but it does seem that the industry has taken it upon itself to appropriate this particular narrative and in the process constructs a particular notion of what might be called contemporary Filipino nationalism.

In this chapter I examine the construction of nationalism within the Philippine call centre industry. Using the narratives of twenty call centre workers, I argue that the industry's construction of nationalism relies on two dominant identity constructions: first, the construction of a particular global identity and a national identity, which are hinged on each other and merge to allow call centre workers to serve the world, thereby also serving the country; and, second, the construction of call centre workers as good, responsible citizens – even heroic, from the point of view of the commercial described above – who contribute significantly to the Philippine economy by virtue of their work in the call centres. Taking these two dominant constructions into account, I argue that the call centre workers whom I interviewed construct a certain kind of cosmopolitan identity; however, this cosmopolitanism is very much grounded in being a Filipino and is construed as a source of Filipino pride. In this sense the cosmopolitan identity hinges on national identity. I also argue that is it crucial to look into the centrality of English in these identity constructions, precisely because it is through the use of English in the industry that these constructions are articulated and enacted. English, in this case, is taken out of its imperialist, anti-nationalist context. It is primarily represented as the language that links the call centre workers to the rest of the world and allows them to participate in it, making possible their service to the world and to the country. As such, it is the language that makes possible their act of nationalism.

These constructions, however, are fraught with tensions and contradictions. For example, the supposedly seamless merging of the global and national identities is not so at all. It is always threatened by callers who challenge or do not recognize the construction of such an identity, largely because the linguistic performance, which is done through the use of English, is deemed problematic. It is also threatened by the negative associations related to call centre work that continue to circulate in the country despite efforts to counter them. Interestingly, when the merging of the global and national identities is threatened, nationalist sentiments seem to become heightened, and the narrative of the call centre workers as modern-day heroes strongly emerges. In light of these tensions the industry's construction of nationalism must therefore be examined, first, within the broader *bagong bayani* discourse in the Philippines and, second, in light of the ideological value of English in the country. Ultimately, an analysis that situates itself within these larger structures of power will reveal that the issues confronting the call centres are in fact the same issues besetting the country in the age of globalization. Thus, while I support the efforts of the industry to construct a positive, desirable image of call centre workers and call centre work, for call centre work is no mean feat, one has to interrogate these efforts within the much wider context of what Blommaert (2005) refers to as the "power effects" of globalization processes.

This chapter is divided into five sections. The first section provides a survey of the existing literature on offshore call centres that focuses on the related issues of identity, culture, and language and the ways in which they are being managed and framed within the industry and, consequently, within the new service economy in the new global order. The second section discusses the link between nationalism and heroism in the context of the *bagong bayani* narrative and how its appropriation in the call centre industry reveals the embeddedness of notions of state and nation in the management of a shifting labour terrain. The third section provides background information on the call centre workers whose narratives form the basis of this chapter's analysis. The fourth section presents the identity-construction efforts narrated by my informants, and their implications for notions of modern-day nationalism vis-à-vis globalization. It is important to note that the identity claims discussed in this section are made not by the state but by the industry itself and the call centre workers themselves. This reveals that the industry and the people in it are very much involved in the way in which they want the industry to be seen, and that they have a stake in the way in which the industry is represented. The last section looks into the tensions and contradictions evident in my informants' identity claims and indicative of wider structures of power.

Identity in the Offshore Call Centres

Much of the available critical work on offshore call centres deals with Indian data (see Cowie, 2007; Mirchandani, 2004, 2008, 2012; Mirchandani & Maitra 2007; Poster 2007; Shome 2006; Sonntag 2009; Srivastava & Theodore 2006; Taylor & Bain 2005, 2006; Vaish 2008). This body of work touches on a number of different issues (for instance, the factors that make offshore call centres possible, and the contradictory demands of call centre work), but one common strand is an examination of the related issues of identity, language, and culture and the ways in which they are negotiated in the offshore call centre workplace. Taylor and Bain (2005), for example, characterize the English-language training in Indian call centres as including practices that can be characterized as "forms of linguistic and cultural imperialism" (p. 274). More specific manifestations of these practices are addressed in Shome (2006), who contends that, in the Indian offshore context, language "functions as an apparatus ... through which the voice of the third world subject is literally erased and reconstructed in the servicing of the global economy" (p. 110).

Meanwhile there are studies that move away from strictly reading the use of English within the framework of linguistic and cultural imperialism. For example, Cowie (2007) looks mainly at the value of accents in the Indian call centres and concludes that, while there is a notion of a standard accent, in reality this so-called standard is generally interpreted and practised differently in different call centres, allowing for possibilities of accents other than the supposedly required American one. Sonntag (2009) provides a comprehensive review of the use of English in the Indian workplace by examining it from the perspective of the various frameworks used in explaining the global spread of English. She argues that, apart from practices that can be considered as instances of linguistic and cultural imperialism, there are those that show resistance to or an appropriation of the hegemonic practice of English in the industry. Vaish (2008) emphasizes these alternative practices by highlighting in her analysis that Indian call centre workers do not necessarily see their work as a threat to their local identity, because they already see themselves as having a "syncretic identity and hybrid linguistic skills" (p. 101). In this regard, Poster (2007) addresses the issue that working in the Indian call centres has become a case of not only doing emotional labour but also "managing national identity," in which workers have to assume different national identities to perform their job.

The most extensive treatment of the subject matter comes from Mirchandani (2004, 2008, 2012) and Mirchandani & Maitra (2007), who provide a detailed discussion of the dominant practices in the Indian call centre workplace – for instance, the kinds of training that the call centre workers undergo, the practice of locational and identity masking, and the performance of both emotional and aesthetic labour. Taking all these practices into account, Mirchandani (2012, p. 1) argues that Indian call centre workers have to perform "authenticity work" in which they "refashion themselves into ideal Indian workers who can expertly provide synchronous, voice-to-voice customer service for clients in the West." They do this by "simultaneously constructing themselves as foreign workers who do not threaten Western jobs, as legitimate colonial subjects who revere the West, as real Indians who form an offshore model workforce providing the cheap immobile labor needed in the West, as flexible workers who are trainable and global, and as workers who are far away yet familiar enough to provide good services to their customers" (p. 9). In the process, Indian call centre workers become American clones to be able to perform their job.

In the case of the Philippines much of the academic work has focused on the discourse features of call centre exchanges between non-native English-speaking call centre workers and native English-speaking customers, and the misunderstandings that arise owing to the cross-cultural nature of these exchanges (Forey & Lockwood, 2007; Friginal, 2007; Lockwood, Price, & Forey, 2008). The body of work is geared towards understanding and explaining the reasons for the communication breakdown between agents and callers, with the goal of improving English instruction and training. The studies definitely provide profound insight into the discourse features of call centre talk, contributing to the work being done on cross-cultural communication and the way in which talk is being managed within globalization. However, these studies are silent on the relationship between call centre talk and larger social processes such as the implications of these new ways of talking in call centres for identity construction in the new global order. Perhaps this more practical thrust is due to the growing anxiety over reports about the increasing decline in the English proficiency levels of Filipinos. The success of the Philippine call centres is, after all, often attributed to the prevalent belief that Filipinos have a high level of proficiency in English (*Philippines: The new outsourcing hub*, 2011).[2]

However, this does not signify an absence of critical work on the Philippine call centre industry. For instance, there are literary productions that confront issues of identity such as the Palanca[3] award–winning play *Welcome to Intelstar* (Martinez, 2006), which was one of the very first in-depth representations, if not the first, of the call centre industry in the country. In the one-act play an English-language trainer, Chelsea, holds an orientation for new applicants of the call centre Intelstar. Although Chelsea is "perfectly bubbly" throughout the orientation, using her "perfectly accented" English, her whole spiel covers those practices that have been characterized in Indian data as demonstrating linguistic and cultural imperialism (for example, locational and identity masking, the accent requirement, and the racial abuse by callers). The incongruity of Chelsea's cheerful demeanor with the alienating practices that she is describing therefore forms the irony that renders the play funny and, more critically, deeply sarcastic and resistant towards not only call centre work but also globalization and its impossible demands on workers. This irony is perfectly captured at the end of the orientation when everyone has left: Chelsea, alone, mutters in Tagalog, "Tang-inang trabaho 'to. Madaling araw na, nag-Iingles pa rin ako" (Fuck this job. It's already the wee hours of the morning, and I'm still speaking English). What the cursing and the return to the local language suggest is the displacement and dislocation that call centre workers undergo in the attempt to perform what, in the Indian data, Mirchandani (2012) calls authenticity work. Tiatco (2014) argues that in *Welcome to Intelstar* call centre work and, for that matter, globalization are represented as a trap, and the Filipino workers as their prey and victims. I highlight this line of criticism here because this is the way in which call centres and call centre workers have been generally represented in Philippine academic circles and in the country's popular imagination (for more examples see Delias, 2014, and Tiatco, 2014). As I will show later in the analysis, it is precisely this kind of representation that has allowed call centre workers to imagine themselves as heroes who suffer the impossible demands of the job to help their families and, consequently, the country.

Within the academe there are several works that attempt to locate agency and resistance in the Filipino call centre workplace (see Bolton, 2010; Delias, 2014; Lomibao, 2007; Salonga, 2010) and also explore the relationship between the use and practice of English in the industry and its implications for identity construction (see Salonga, 2010). Lomibao (2007), for instance, notes that Filipino call centre workers feel "deterritorialized," "dislocated," and "alienated" owing to the night-time shifts, the irate customers, the language and accent requirement, and the high-tech monitoring system; nevertheless, they are also able to "[employ] various strategies of resistance to ease their dislocations: they [drop] calls, put customers on hold for long periods of time, [refuse] to speak English despite strict policies, [deviate] from their scripts, and [answer] back at callers" (vi). Bolton (2010) investigates the way in which the "rhetoric of globalization gives way to a consideration of lives lived locally" (p. 563), by drawing on the linguistic experiences of Filipino call centre workers, and he concludes that transformative linguistic practices also exist in the industry. Delias (2014) critiques the representation of call centre workers as desiring an American identity, arguing instead that the workers are able to maintain their local identities and are in fact proud to be not in the United States (also in Tiatco, 2014). My own work on the Philippine call centres (see Salonga, 2010) from which the interview data in this chapter are taken is an exploration of the way in which the dominant linguistic and communication practices in the industry provide possibilities for situated agency primarily because of the alternative meanings that call centre workers assign to these practices. In this chapter I follow the same critical tradition in which the works reviewed above are situated. However, I focus particularly on the intersection between nationalism, identity, and language, and the ways in which this complex relationship is understood, negotiated, and contested by those on the ground, the Filipino call centre workers themselves, vis-à-vis globalization processes.

The *Bagong Bayani* in Virtual Migration

The term *bagong bayani* was first used by former president Corazon Aquino "in a speech she gave in 1988 to a group of domestic helpers in Hongkong, telling them that 'kayo po ang mga bagong bayani' (you are the new heroes)" (Rafael, 1997). In 1989 the state officially established the Bagong Bayani Awards, creating "a national search for the country's outstanding and exemplary Overseas Filipino Workers (OFWs). The awards seek to recognize and pay tribute to our OFWs for their significant efforts in fostering goodwill among peoples of the world, enhancing and promoting the image of the Filipino as a competent, responsible and dignified worker, and for greatly contributing to the socio-economic development of their communities and our country as a whole" (Bayani, 2014).

From the wording of the awards it is apparent that the OFWs are constructed as playing a significant role both globally and locally. Globally, they become almost ambassadors of the country, "fostering goodwill" and "promoting the image of the Filipino." To do this, however, the OFWs endure many hardships: leaving the country, separating from their families, and suffering a variety of abuses in foreign lands. But this is precisely how they are able to support their families and, through their remittances, to help keep afloat the country's economy. This is also the reason that locally they are made to occupy the role of "heroes."

The idea of great sacrifice is therefore the foundation that underlies the narrative of the *bagong bayani* or the new or modern-day hero. According to Roxas-Tope (1998), the Filipinos' concept of nation and nationalism is rooted in the idea of martyrdom or sacrifice, as is clearly suggested in the last line of the Philippine national anthem: "Ang mamatay ng dahil sa iyo (I am willing to die for you)" (1998, pp.13–14). The notions of nationalism and heroism in the context of the Philippines are connected to the idea of sacrifice, of suffering even death, for love of family and country. The heroes of past Philippine revolutions suffered and died for this love. The struggles of the OFWs take a different permutation from those of past heroes, but highlighted in the narrative is the idea that, like the country's national heroes, they will suffer anything for their loved ones and their country and are thus heroes too.

The "Pedestal" commercial indicates that the call centre industry seems to be appropriating this narrative. As discussed above, call centre workers often experience alienation, dislocation, and a variety of mental and physical health issues as a direct result of their job, mainly because of the hours that they keep, the effort that authenticity work entails, and the racial slurs and abuse that they receive from callers (see Delias, 2014; Lomibao, 2007; Mirchandani, 2012; Salonga, 2010). As also mentioned earlier, in some sections of Philippine society call centre workers are not held in high regard. They are seen as glorified telephone operators, a perception probably influenced by early criticism of call centre work that depicts call centre workers as little more than well-scripted robots that say the same things in the same way, with little or no control over workplace practices and their own linguistic production, and with no possibility of career advancement (see Cameron, 2000a, 2000b). Further, they are judged as pretentious and superficial, much like the English that they use at work is also often evaluated, and as having no other concerns but drinking, partying, and purchasing the latest fashion and gadgets. These perceptions suggest that call centre workers face great struggles at work (from both the foreign callers and, sadly, fellow Filipinos), which they, like the OFWs, bear and suffer for love of family and country. The *bagong bayani* narrative applies in this regard.

Using Indian data, some scholars have highlighted the similarities between call centre workers and migrants or migrant workers by describing call centre work as a form of "virtual migration" (Aneesh, 2006; Shome, 2006). Shome (2006) asserts that virtual migration takes place when "the body both departs into another national world and time in the performances of American-ness through virtual technology but also simultaneously remains geographically and temporally situated in India" (p. 116). This is true as well for Filipino call centre workers in that, as they perform different identities for different sets of callers in different virtual spaces, their bodies nevertheless remain very much contained in their offices in Manila. The notion of virtual migration takes on further significance in light of the fact that the state-authored narrative of the *bagong bayani* seems to have been appropriated by the industry as a tool to recast its image in terms that are desirable and attractive, precisely to counter disparaging ideas about the job.

In this chapter I explore the way in which the *bagong bayani* narrative is made to traverse this rather uncharted terrain of virtual migration. As the analysis will show, there does not seem to be an explicit acknowledgment that the *bagong bayani* narrative is being appropriated, so it is difficult to assess whether the call centre workers see themselves as occupying the same position as OFWs, or if they are at all aware that their claims of heroism sound very similar to those found in the OFW-as-hero narrative. It is also possible that because the construction of the call centre worker as hero is a very recent development, a move that was first made in 2013 by Convergys in what the company bills as the very first media campaign in the country acknowledging the "heroic efforts" of call centre workers (Convergys, 2013b), further inquiry needs to be made into the matter. Nevertheless, I aim to demonstrate that the *bagong bayani* narrative provides a productive lens with which to examine the identity claims of the call centre workers included in the study as it reveals that, even in the virtual and supposedly borderless spaces occupied by call centre workers, there remains a need to invoke country and nation in the construction of identity. In short, even if the physical bodies of call centre workers do not move, it is clear that they carry their materiality with them in their transactions with callers, influencing the way they see and present themselves to the world. Further, using the *bagong bayani* narrative as lens to examine call centre work reveals that the issues confronted by call centre workers are not unique but parallel those of other types of workers in the new service economy.

The Filipino Call Centre Workers

This chapter draws primarily on interview data (treated as narratives) to surface the identity constructions that are being made in the Philippine call centres. I conducted the interviews from June 2007 to September 2008 as part of a larger body of work on sociolinguistic practices in call centres and their implications for identity construction and linguistic agency (see Salonga, 2010; see also Salonga, 2015, and Tupas & Salonga, 2016, for other implications of the interview data). I recruited the informants through the social network (or friends of friends) method (Milroy, 1980) and carried out the interviews in a semi-structured manner with a set of guide questions that focused on two main topics: the speech style used in the call centres, and the perceived value of the speech style. As the interview data will show, while the questions were not specifically about nationalism or heroism or the sacrifices of call centre workers, these notions came up often in relation to the perceived value of call centre work. The issue of English and its important role in performing call centre work also came up frequently. These responses thus form the basis of the analysis that follows in the next section.

I interviewed twenty call centre workers. The majority of my informants are front-line call centre workers – or, more popularly, "agents" or "customer service representatives" (CSRs) – and English-language trainers and specialists from different call centres. These call centres service different parts of the world, but the majority of the customers come from the United States. For this reason my informants are exposed to a range of accents and also try to employ different accents in their performance of the job. The age range of the informants is between twenty-one and thirty-seven years old, with the majority being in their early- to mid-twenties. Their monthly pay ranges from US$250 to US$400 for front-line customer service work, and the English-language trainers and specialists earn up to US$600. They work in nine-hour shifts including an hour of lunch and thirty minutes of short breaks. The informants come from a wide range of educational backgrounds. All of them have a bachelor's degree (in English, dentistry, management, tourism, and chemistry, among others) except for one, who has a high-school diploma. They come from different schools, some top-tiered (for example, the University of the Philippines and De La Salle University) and others not as highly ranked (for example, Mother of Perpetual Help Institute or AMA University and Computer Colleges). Of the twenty interviewed, twelve are male and eight are female. This profile provides a generally accurate representation of the industry. The Philippine call centres are populated by young people from different educational backgrounds who perform different customer service tasks for a wide customer base. However, the gender distribution is not entirely accurate. I have two more male than female informants, but the Philippine call centre industry employs a greater number of female than male workers ("2010 Annual Survey of Philippine Business and Industry").[4]

The use of interview data brings into focus the complex and conflicted relationships that call centre workers have with the kinds of identity constructions that they have to make to perform the job. As Heller (2011) notes, interviews are "important for getting a sense of participants' life trajectories and social positioning ... They are also important sources of accounts, which allow glimpses into the beliefs and values and ideologies that inform what people do and why they do it" (pp. 44–5). The interview data highlight the way in which anxieties and tensions among call centre workers reflect wider relations and hierarchies of power, on both the local and the global levels. Freeman (2001) argues that "not only do global processes enact themselves on local ground but local processes and small scale actors might be seen as the very fabric of globalization" (pp. 1008–9). Nevertheless, I recognize that interviews are performances as well: "they are what a certain kind of person tells another certain kind of person, in certain ways, under certain conditions" (Heller, 2011, p. 44). The sentiments in the interviews are therefore regarded not as conclusive but rather as suggestive of certain patterns and structures that have emerged in the call centre workplace and are positioned both locally and globally.

The other set of data that I use to supplement the interviews consists of three videos created by Convergys. One of these videos is the "Pedestal commercial" described in the introduction, which ran on local television networks in 2013. Two videos, one titled *Heroes Within* (Convergys, 2013a) and the other *A Legacy of Heroes* (Convergys, 2013b), were released only on YouTube. The choice to include these videos came about when I saw the "Pedestal" commercial on television. When I went on YouTube to view it, I found the two *Heroes* videos there. All the videos revolve around the same theme – call centre workers as heroes in light of the significant contribution of the industry to the Philippine economy. Given this content, the videos function to historicize Convergys's presence in the country and to promote the company in order to attract prospective workers and lift the image of the industry in the local context. Convergys is a global call centre with "125,000 employees working in more than 150 service centres and 31 countries" ("Welcome to Convergys"). It was one of the first call centres to establish a site in the Philippines, and it is currently the biggest with about 35,000 employees in the country (Convergys, 2013b; O'Malley, 2013). Convergys is therefore influential in the creation of the image that the industry wants for its workers. For this reason the Convergys videos provide useful insight into the management of the call centre worker's identity in the industry and, in comparison to the interview data, the extent to which call centre workers themselves support this effort.

Finally, supplementary data, including recent newspaper reports, informal conversations with industry friends who are language trainers and specialists, and insights gathered in call centre conferences attended and call centres observed,[5] are also used when necessary to further extrapolate some points gleaned from the first two sets of data.

Voices from the Philippines

In the analysis that follows I draw on the assumption already established in the above literature review: the offshore call centres are a productive site for examining the management and negotiation of identity within globalization processes. With the inevitable meeting of the local and the global at the very heart of offshore call centre operations, identity in this case does not refer solely to individual constructions of the self. Rather, these individual self-presentations provide tremendous insight into the way in which notions of nation and nationalism are being imagined in the new global order. This means that mundane everyday practices such as going to work and performing one's job and talking about it in particular ways have something to say about a nation's projection of itself in relation to other nations. In this sense, the analysis follows from Billig's (1995) notion of banal nationalism to refer to the "ideological habits" that "are not removed from everyday life" which make it possible for the idea of nation and nationhood to be established (p. 6). The analysis assumes that nationalism is "reproduced in a banally mundane way," in that "[daily], the nation is indicated, or 'flagged,' in the lives of its citizenry" (Billig, 1995, p. 6). These ideas about nation and nationalism are then drawn out of the narratives of the call centre workers and then analysed using contemporary understandings of such concepts.

Global and National Identity Constructions

In this section I look into the way that the call centre workers whom I interviewed perceive their work in the industry as a performance of a global identity. They do this by (1) seeing their work as a link to the global community, (2) showing appreciation for different cultures, and (3) highlighting that they have a significant role in the new global order. First, they see their work as linked to the global community because it enables them to communicate with people from different parts of the world. Luis, a quality assurance manager (formerly a CSR),[6] for instance, talks about how the call centre industry provides workers with exposure to the world. Accordingly, in being exposed to the world, they are now able to move and participate in it:

> You know, our reps [CSRs], they get to hear people, Indians, Norwegians, Irish, Brits, Mexican-Americans, so they can easily identify things, because we're so used to having different sets of vocabularies, words, systems. So outside of the hours of call centres, for example, you go to another place, perhaps a particular country, you have less difficulty to be misunderstood. For example, India, you wanna ask them things, you understand them. If you go to the UK, you would understand them because you're used to the accent. You can go to Scotland and you understand them ... Being used to having this very adaptable style, you can deal with anyone, whatever race, wherever from.

In Tupas and Salonga (2016, p. 6) this sentiment is articulated by Lloyd, a CSR, who talks about his delight in being able to communicate with people coming from other parts of the world by using different varieties of English and about his pride in being acknowledged as occupying different identities precisely because of how he uses English:

> What I appreciate is that ... I get to listen to a lot of different variations of the accent. The interesting part is talking with Irish and Scottish people. It's really different, the way they say words. It's like our style of pronouncing each syllable of the words, but they still have the English accent though, which is very different. I find it very interesting. It is actually fun.
>
> Sometimes I play with my accent, I use [a] British [accent], then the customer would say, "Are you from Australia?" ... "No, I'm from the

Philippines." "I thought you're from Australia 'cause you sound Australian." Then there's this customer who feels comfortable when he's talking to me because he [thinks] I'm an Englishman and he says, "You know, my father's also English."

From Luis's and Lloyd's accounts we see that call centre work is not only about being connected to the global community; it is also about inhabiting an identity that is globally recognizable and intelligible to the callers. English is crucial in this identity construction because it is the language that enables this global identity.

Second, for a number of my informants, the call centre workplace has given them the confidence to communicate with all kinds of people. Jean, a CSR, recounts that talking to foreigners on the telephone all the time has helped her to overcome her fear of talking to them:

Dati meron akong, di ba some people have, this thing about foreigners, you don't talk to them. "Ay 'wag kang titingin dyan, baka ka kausapin nyan." E ngayon, if you hear their accents all the time, you talk to foreigners all the time, foreign bosses ganun, doesn't matter anymore.

(I used to have this thing about foreigners. "Hey, don't look there, there's a foreigner. He or she might talk to you." But now if you hear their accents all the time, talk to foreigners and foreign bosses all the time, it doesn't matter anymore.)[7]

Janice, also a CSR, says that talking to different kinds of people has made her grow as a person because it has allowed her to get to know them: "Dealing with these people every day, being able to understand them, whatever accent they have, because we speak with Latinos, we speak with black Americans, we speak with rednecks,[8] being able to understand these people and address their issue at the same time and being friends with them over the phone in just a short period of time, explains or tells a lot about how you have grown as a person." She continues by saying that her work in the industry has expanded her horizons: "It's like I have experienced another world. It's not like I'm boxed. I've been living here in the Philippines, but it feels like I have already travelled."

Karen, an operations manager (formerly a CSR), shares the same sentiment: "I would think it has broadened my perspective of what's going on in other countries. I am a mother and I'm a married person [but]

I have other interests outside of being a mother or being a wife. The discussion has broadened. [It's] more political."

The extracts above suggest that being able to communicate with different kinds of people is the first step. Ultimately, talking to people from different parts of the world provides a change in perspective and a sense of becoming a better person, because one becomes more appreciative of and sensitive to cultural differences. As Jean notes:

> Parang more or less meron ka nang, ah, 'pag ganitong client, ito ang important sa kultura nila, so you get to know a bit more about other cultures, kasi you never, never ask someone from this culture na ganito ... Saka mas may care kang baka maka- offend ka. Ano ba, sa culture ba nila, ganito? Nagiging sensitive ka, be careful of what you say.

> (You develop this kind of sensibility. When you are talking to certain clients, you learn what's important in their culture, so you also get to know more about other cultures. Then you learn that there are certain things that you cannot ask or discuss when you're talking to clients who come from this culture. You become more careful that you don't offend them. You really just become more sensitive, more careful of what you say.)

Third, because of this heightened understanding and appreciation of the world, my informants envision themselves as having a place in the new global order. They see themselves as helping, and in some cases even providing information and knowledge, as they resolve the everyday problems of their callers. For the majority of my informants, too, this act of helping or "educating" is not just lip service; it is felt sincerely. Lloyd, for instance, talks about going the extra mile for customers: "As much as possible we also go the extra mile of resolving the issue. Even if it's gonna take so much of our time ... because we don't want the customer to use another step in resolving the issue."

Janice adds that this is not just about assisting the callers but also about "educating" them: "We have to educate the customer while doing the troubleshooting. It's not just do this and do that. We don't do that. We just don't tell the customers what to do. We tell them why we do a certain step. And then, in coming up with a resolution, we also provide explanation why a certain issue in the first place occurred and why it's happening." In the end, Sarah, a CSR training to be a language trainer, says that there is real fulfilment in being able to resolve the callers' issues: "You want to give them things that can solve [their problems] so that, you know, at the end of that call there's fulfilment. You wanna take the next call and just solve everyone else's problem."

It is apparent in these accounts that employees see the work that they do as significant. They are, after all, serving other people – even beyond the bounds of the call centre and their formal job descriptions. Interestingly, English also has a critical role in this regard. Josh, a language trainer (formerly a CSR), says: "A lot of the Korean tutors now for English are call centre agents. A lot of them. It's their sideline, because 'I can speak English now, my English is better now, why not share it to the world?'" Here Josh is talking about Koreans whose online English tutors are actually call centre agents who take tutoring jobs on the side. Again, the use of English is a crucial part of this global identity because English becomes the conduit for its construction. For Josh, English is the language used by the rest of the world.

It is important to note at this point that the orientation towards American English or the construction of an American identity seems to have lessened in the Philippine data. There seems to be a move away from English that is considered to be "too American." Instead, there is a criss-crossing of various identities through the use of the different varieties of English. My informants associate this trend with several developments in the industry. Will, a language trainer (formerly a CSR), links it to multiculturalism within certain countries where the customers are located: "Some call centres don't want to be labelled as 'it's too American.' They don't want to be labelled like that. They would always want to say that they are not only entertaining the Americans because they are aware that in the United States it's not just Americans. It's actually a melting pot. There are so many cultures and there are so many people there in the U.S. And that you should have to be able to address everything" (also cited in Tupas and Salonga, 2016, p. 7). Lloyd explains that this may also have to do with the increasing number of countries that Philippine call centres are serving: "Because we're catering to the global needs, we don't only take calls from the U.S. It's actually worldwide. You can speak with people with thick British accents, Welsh accents, with Arabic accents, with Indian accents, so we adjust in that particular aspect" (also cited in Tupas and Salonga, 2016, p. 7). As call centre workers disassociate from an English that is "too American," they lean towards Englishes that signal multiple identities.

Sonntag's (2009) notion of linguistic cosmopolitanism offers some assistance here. It signals the way in which individuals are freed from the constraints of their own local contexts. This freedom extends to linguistic experience, where "language is created dialogically. Linguistic cosmopolitanism embraces a fragmentation of English, a diversity of Englishes" (p. 15). Such fragmentation may be likened to what Vaish (2008, p. 99) has characterized as "a syncretic pastiche of multiple linguistic identities," something that Sonntag has observed in her own data as well (2009, p. 17).

My informants incorporate many elements of this cosmopolitan identity. They adopt an orientation that is global and connected to the world; open to and appreciative of other cultures; and crisscrossing into different identities through a knowledge of, and ability to use, different varieties of English. However, their identity departs from the cosmopolitanism of Sonntag and Vaish in several key ways. It is not free from the local context, and it does not disavow national affiliation. The cosmopolitanism of the Filipino call centre worker remains very much grounded in being a Filipino. It is constructed as a source of Filipino pride, primarily because call centre workers see their global-ness as resting on their Filipino-ness (that is, enacting the so-called naturally Filipino work skills), and because, against the backdrop of negative perceptions surrounding call centre work, employees recast their work as a form of sacrifice to fend for their families and help their country. Coming full circle then, we see how the message of the commercial in the introduction to this chapter plays a role in the identity construction and sense of self for these workers and how the *bagong bayani* narrative may apply to their own experience.

Enactments of Nationalism

In this section I explore two supposedly "natural" Filipino characteristics, highlighted by call centre workers, that make them perfect for call centre work in particular and as global citizens in general. The first one is the belief that Filipinos are naturally caring and nurturing, and the second is the idea that Filipinos have the gift of language, which means that they can easily adapt their speech to suit the particular demands of any communicative situation. In talking about the first characteristic, my informants refer to the fact that Filipinos have a natural concern for others. Christian, an operations manager (formerly a CSR), for instance, narrates the account of an Indian who established a call centre business in the Philippines. According to Christian, the Indian says that the uniqueness of Filipinos is their *"malasakit* [concern for others]. I can't even find an English word to describe what *malasakit* means. People here in the Philippines will generally *malasakit*. I won't be able to fully understand what *malasakit* is, ... because I'm not a Filipino, but that's a very, very powerful verb [it is actually an adjective]. That's one thing that I can verify because most of the Filipinos I've worked with worked as if they owned the company." This *malasakit* is not just reflected in one's behaviour or action, but it actually manifests in the voice of Filipinos. Charles, a CSR, states that "we Filipinos have this type of *malambing* [sweet, caring] tone. We speak in a way that you are trying to tell the customer that you're willing to help. Because primarily it's the reason why you're working. You exist to help the customers over the phone."

Others, aside from my informants, have referred to this particular trait. The Business Processing Association of the Philippines (2008), the official umbrella organization of the nation's business process outsourcing industry (of which the call centre industry is a part), characterizes the Filipino workforce as possessing "interpersonal warmth." This is echoed in the website of the Contact Center Association of the Philippines (2013), the umbrella organization of the Philippine contact centre industry, which describes Filipinos as globally renowned for their "friendly attitude and innate warmth as a people."[9]

The second supposedly Filipino trait is a certain facility with language – to project an English that is neutral and therefore intelligible to callers. Interestingly, my informants compare their neutral English to the unintelligible one of Indian call centre workers. Janice, for instance, says: "That's one selling point of Filipinos compared to Indians. You just really need to decipher what they [the Indians] are saying because it's not clear, the accent. It's not clear, you can't understand it. Filipinos, we talk clearly, that's what's important." Andy, a language trainer (formerly a CSR), extends this further. He says that the English of Filipino call centre workers is so neutral that there are no traces of Filipino-ness. He then represents this facility as a gift of language that is "natural" to Filipinos: "It's better to talk to a Filipino than to an Indian. The Filipinos are gifted. Like for example when you hear a Filipino talking to another caller, you will not notice that that person is Filipino. We have the gift of the tongue. We have the gift of language. The Philippines [has] the gift of language, and that's proven already."

As already mentioned, the kind of cosmopolitanism experienced by my informants is very much rooted in being "a Filipino." It conflates both global and national identities: both seem to derive value from each other. In the accounts above, this global identity works for employees by drawing upon "the essence" of a Filipino. A call centre worker's Filipino-ness has currency in the global market. Yet at the same time it is the very absence of Filipino-ness that makes the identity particularly desirable and attractive. English is central to this construction because Filipinos are constructed as being able to employ an English that does not index any local-ness, allowing them to successfully perform the job of global call centre workers and move in the world.

These identity claims are not always recognized or accepted by customers. While my informants contend that the English of Filipinos is neutral, they have anxiety about the way callers perceive their English. Language is crucial for them because if their English is not considered legitimate (or, in Mirchandani's term, "authentic," 2012), neither is the identity that is attached to it. Lloyd, for instance, talks about "customers who would make you feel like you're not deserving [of] speaking that language. 'Cause I'm not a native speaker, sometimes I commit lapses, and sometimes it really makes you feel that you are different from them [customers]. They get to that point that they really had to say it, 'Okay, I'll repeat.' There are those markers that [are] really belittling of your personality" (also cited in Tupas and Salonga, 2016, p. 10). In some cases the abuse is straightforward, as Luis relates: "Some would say, 'I'm angry at you 'cause you're a Filipino.'"

Sadly, this mistreatment of workers comes from outside the call centres as well. As established earlier, negative associations about call centre work circulate freely in the country. A friend of mine who works in the call centres recounted that employees often do not tell people about their jobs for fear of being judged or looked down upon. For one thing, people think of call centre work as easy, and therefore belittle the job. Lloyd, for instance, notes: "You have to adapt [to] the usual stereotype that they think about call centre agents – that it's so easy to get a job in the call centre. It's a stereotype, it's a prejudice, but it's not really true. It's not really easy at all."

Another negative impression is that call centre workers do not have any real concerns. They are young and have purchasing power, and all they do is socialize. Eric, a CSR, talks about his impression before entering the industry: "Mostly, my impression was that people who work there, all that they do is to party and to drink, to drink coffee, smoke, and just always party and party."

People also criticize the English of the industry as artificial and discourage its use outside of the context of call centre work. Wendy, a CSR, says that she and another friend, who also works in call centres, are very careful about the way they speak with non-call centre friends.

If they start talking with an "accent," their friends really show their displeasure:

> Kasi kunyari may mga friends ako, tapos dalawa lang kami na nag-wowork sa call centre nun, tapos kapag nag-uusap kami, tapos parang maglagay ka lang ng konting accent na ganyan, parang "hay naku, usapang call centre na naman to."

> (When my friend who also works in the call centres and I are with our non-call centre friends, and we start speaking with a little bit of accent, our other friends would start saying, "Oh, they're going to start talking about call centre work again.")

In the face of all these negative perceptions, coming from both callers and fellow Filipinos, it is not difficult to imagine that workers feel they are "under attack." This is in fact the term used by one of my call centre friends to describe the feeling of working in the industry.

However, it is in the moments described above, when their efforts at identity construction are challenged, that call centre workers display a heightened sense of nationalism. One way in which they respond to the attack is by positioning themselves as better than the callers whom they are trying to serve, an observation that is true in the Indian data as well (Poster, 2007). They label the English of customers as substandard, and their own as more valid and legitimate. As non-native-English speakers, these call centre workers see themselves as more adept users of the language than are the customers. Moreover, in the midst of deciphering the (sometimes poor) English of the callers they are also solving the customers' problems. Lloyd points out: "Being a non-native speaker, it also makes [me] think that I'm still a better speaker than this person is. I have office mates who criticize some other callers because they can't speak English well. It gives a sort of opposition with being superior to them because we know the language and they don't and they need our help. There you have two barriers already, the knowledge of the problem and the knowledge of the language."

My informants say that, even though customers ridicule them, they know that they are doing something good. They are helping the callers; they are helping themselves and their families; and lastly, they are helping the country's economy. They see themselves as responsible citizens who may be doing more for the country than are the people who belittle them. As Christian notes, "call centre work makes them [CSRs] better individuals. [They become] more successful so they can help their families better. They can help the country better in the longer term, and they can be responsible citizens in the longer term, but in a year's time they can make their lives better."

This sentiment is made very clear in an incident involving a local television show, *The Borrowed Wife*. Two characters in the first episode exchanged the following lines when talking about the prospect of applying for a call centre job: "Hindi ako nag-aaral para sumagot lang ng telepono!" (I did not go to school just to answer phone calls!), and "Pang walang pinag-aralan lang 'yan" (That job's only for uneducated people). An overwhelming number of furious responses came from the industry (De Jesus, 2014). One group of workers vowed to boycott the show. Some expressed their anger in a podcast: "The gall to belittle an industry with billions in revenues making it one of the chief economic drivers of this country. An industry that employs hundreds of thousands of hardworking people who pay billions in taxes to the government. What about them? How much do they know? How much research have they done to make such sweeping, degrading statements?" (Cruz, in De Jesus, 2014). Clearly, call centre workers have a stake in how they are perceived and represented in mainstream media.

Tensions and Contradictions

As apparent in the preceding section, there are tensions and contradictions in the identity claims of my informants. Call centre workers are aware of these tensions – thus the continuous reconstruction and repositioning of their identities. In this section I hope to show that the ongoing identity construction (and reconstruction) taking place in the industry is indicative of larger structural problems. First, the analysis illustrates that the depiction of call centre workers as heroes is in fact an appropriation of the *bagong bayani* narrative. However, the *bagong bayani* narrative is problematic. Rafael (1997) notes that it functions to "contain the anxieties attendant upon the flow of migrant labor, including the emotional distress over the separation of families and the everyday exploitation of migrants by job contractors, travel agents and foreign employers." In other words, the narrative works to legitimize the state-sponsored "exporting" of Filipinos. It hides the state's inability to provide jobs for its citizens within its own borders, and its failure to protect them when they work abroad (Cheng, 2010; Guevarra, 2010; Rafael, 1997; Rodriguez, 2002; Rupert & Solomon, 2006). The narrative also romanticizes the notion of sacrifice. It makes the act of suffering noble, heroic, attractive, and inevitable. Yet in reality the plight of OFWs is extremely difficult. In addition, having assumed the role of labour broker, the state is partly responsible for these structural inequalities in the new world order (Lorente, 2007, 2012). The borrowing of the narrative into call centre work carries the same problem. In legitimizing and acknowledging the work of call centre workers as heroic, the industry also masks the extremely difficult position in which many call centre workers find themselves. While call centre workers may not be physically harmed by the callers, the verbal abuse they receive from racist and irate customers is not any less harmful to their sense of well-being.

Second, the depiction of Filipinos as caring and nurturing is not necessarily new. Philippine labour agencies have capitalized on the notion of "able and caring hands" in promoting Filipinos as workers of the world (Lorente, 2012). Even the Philippine tourism industry has capitalized on the supposedly genuine smile of Filipinos in attracting tourists to visit the country. The Department of Tourism's (2009) website, for instance, highlights the legendary hospitality of Filipinos, especially towards Western visitors. The construction of Filipinos as naturally caring therefore colludes with the *bagong bayani* construction in bringing about further exploitation. It suggests that Filipinos should be able to take whatever is thrown at them because they care and really want to help – never mind that they are being abused in the process. The construction of Filipino call centre workers as "naturally" caring puts the workers in a double bind. On the one hand, it makes them attractive and desirable workers, but, on the other hand, it allows them to be exploited by a system that already tells them not to answer back even when the callers are rude, insulting, or racist. It can be argued that the construction of Filipinos as *mapagmalasakit* (compassionate and caring) and *malambing* (sweet and thoughtful) is a means of cultivating a more palatable form of subservience.

Finally, there is the issue of English. There has been a move beyond American-ness, and therefore there is a wider range of identities for call centre workers to inhabit. This is a good thing in the sense that workers do not necessarily have to perform one national identity; they can be other kinds of people too. In fact, they can "be" Filipinos. Most of my informants note that, because outsourcing is now a widely known practice, they no longer have to hide their location or identity. When asked by customers if they are Filipino, they can affirm it. When asked where they are, they can say that they are in Manila. With many identities, these workers are engaging multiple varieties of English, both native and non-native. This pluralism may prove advantageous to call centre workers, who will then have greater linguistic resources at their disposal, even outside the workplace.

However, this managerial practice also remains problematic. It does not totally erase the requirement for authenticity (Mirchandani, 2012). Even if workers are not required to engage in locational or identity masking, they are subject to other kinds of authenticity work. This includes speaking in a form of English that is perfectly intelligible to the caller. The caller's language and identity, in this case, still dictate the form of English to be used by the call centre worker. Accordingly it creates more impossible demands on the workers. As there are different kinds of callers, coming from different parts of the world, workers are now expected to use many varieties of English as they reflect and display those different kinds of identities. Workers must conform to the requirements of service in the age of neoliberal globalization: being flexible, adaptable, and consistent throughout – the perfect worker.

In the narratives of my informants English is constructed as a means to connect with the rest of the world and as a means to foster certain identities that are globally recognized and recognizable. English, ironically enough, also facilitates their acts of nationalism as Filipinos. These are conflicting constructs. English is considered, in one sense, for its pragmatic purpose (that is, to communicate and do one's job) and simultaneously for its interpersonal purpose (that is, to index particular kinds of social identities). English is the language of non-Filipino-ness; yet simultaneously it is precisely this ability to speak English without giving away any trace of Filipino-ness that marks the call centre worker as appropriately "Filipino." Sociolinguists note that languages often carry different kinds of meanings and values at the same time, depending on the people who use them and the kinds of contexts in which they find themselves (see Blommaert, 2003, 2005, 2010; Pennycook, 2001), so these conflicting constructs may not necessarily be problematic; however, they become so when these contradictory meanings are not recognized. In the case of the call centre workers, English may mean various things to them. It may signify multiple identities depending on the way it is used, but for the most part it is a tool that they use to perform their job. For callers, however, it is possible that English is not simply a tool but a marker of a fixed cultural identity, such that they expect a particular variety of English because they see this as indexing a particular identity; any other variety is not acceptable.

Even without the callers to influence the way in which the workers use English, the Filipinos' relationship with English is already one of ambivalence and conflict given the Philippines' colonial history (see Tupas, 2001, 2004, 2007). Within the country English continues to serve a stratifying and gatekeeping function: it is a marker of social class and a core foundation of social hierarchies (Tupas, 2001, 2004, 2007). This is true also in the call centres. Workers who are able to play with different varieties of English and thus cross into different identities are usually the ones whose English proficiency is high. These workers tend to come from the middle class, have access to technology and global media, and have attended private schools. Applicants with below-par English proficiency tend to struggle in the early stages of the recruiting process and may not even break the employment barriers to enter the industry (Salonga, 2015; Tupas & Salonga, 2016). Reports, for instance, have shown that only about 5 per cent of the total number of call centre applicants (roughly 400–500 per week) are hired, largely owing to poor English skills (Forey & Lockwood, 2007). The English proficiencies of these applicants are definitely not equal (Salonga, 2015; Tupas & Salonga, 2016). In this way English plays a similar sorting and stacking role within the call centre industry as it does in the wider Philippine society. The asymmetrical relations of power that constitute the use of English in the Philippines and in the call centres make it difficult to equalize the kinds of social meanings and identities that people will attribute to the language. In the end, while my informants can assign their own meanings to their use of English in the industry to empower themselves, this can be closed off by the more dominant ideologies about English that circulate in the world (Tupas & Salonga, 2016).

Conclusion

My main goal in this chapter has been to examine the ways in which the intricate relationship between nation, identity, and language is managed, negotiated, and comprehended in the Philippine call centre industry and creates a particular form of contemporary Filipino nationalism. To achieve this goal I analysed a commercial released by Convergys, one of the biggest call centres in the Philippines today, and the narratives of twenty call centre workers, following Billig's (1995) notion of banal nationalism. The assumption is that in the mundane, everyday ways that call centre workers talk about their work and its value it is possible to surface ideas about nation, nationalism, and national identity that can provide a better understanding of the way these concepts are being carved out in the age of globalization.

My analysis reveals that the industry and call centre workers actively participate in the construction of these concepts as they themselves claim a certain cosmopolitan identity that is distinctly Filipino. The construction of this identity is based on a global identity, on the one hand, and a local identity, on the other. The global identity is one that is connected to the rest of the world and is open and appreciative of other cultures; it is an identity that has a role in the new global order – that of serving the world. This global identity, however, is hinged on the local identity, as it draws meaning primarily on its Filipino-ness and on the way in which, in serving the world, it is able to perform its local role of helping the country. Performing the global identity thus entails the performance of the local identity as well, and vice versa.

The use of English is critical in the performance of this identity, for it is through English that the call centre workers see themselves as able to serve the world. In serving the world, they are able to serve the country as well; English then becomes the language that makes possible their act of nationalism. This act of nationalism is strengthened by the Convergys commercial's representation of call centre workers as heroes who are in large part responsible for the continued growth of the Philippine economy. As such, these heroic efforts of call centre workers should be acknowledged and respected. This move of Convergys, while new in the context of the call centre industry, in fact parallels the *bagong bayani* narrative that the Philippine government has used to celebrate overseas Filipino workers.

However, more crucial are the tensions and contradictions revealed by the analysis. The analysis shows the complex and complicated ways in which the industry and the call centre workers themselves make sense of their work and their value both locally and globally. It is clear that the workers have a stake in the industry and actively participate in the way it is represented and imagined. However, it is also clear that their efforts at identity construction are implicated in wider structures of power within globalization. Claiming a seamless global and local identity integration, for instance, seems to be a way of masking the inherent contradictions in call centre work and the impossible demands of the new service economy. The idea of a neutral, unmarked English that one is able to wield at any moment to communicate with any person from any part of the world seems to be a response to neoliberal globalization's idea of the perfect worker: flexible, adaptable, and yet consistent. Even the reframing of all these as efforts directed towards helping one's country, while commendable, suggests that the call centre workers' sense of self and value is caught up in the politics of call centre work in the country and in the world.

This is not to say that the call centre workers do not have control over their identities and their representation of them, because they themselves challenge and continuously negotiate these identities. Thanks to this active intervention, the everyday discourses in which call centre workers engage – among themselves, with callers, with friends (and foes) outside of the industry – may result in unexpected transformations. Indeed, the identity claims of the call centre workers should be seen within broader structures of power, but also within moments of individual expressions of agency. In the end these everyday negotiations and contestations made by those on the ground, by those who are on the front-line and directly involved in the minute details of globalization processes, have the potential to change existing practices and landscapes. It is, after all, in the banal and commonplace articulations of identities that notions of nation and nationalism are crafted.

NOTES

1 It is possible that the different scenes in the "Pedestal" commercial depict the different locations in which Convergys has opened an office or a site.
2 The reference to the Filipinos' ability to speak English well is also evident on many websites promoting the Philippines as a call centre destination (for example, see www.ccaponline.org, the official website of the Contact Center Association of the Philippines).
3 The Carlos Palanca Memorial Award is the most prestigious literary award in the Philippines.
4 This is based on the latest statistics issued by the Philippine National Statistics Office (NSO). The gender distinction is not salient in this chapter. For an exploration of the gender dimensions, see Salonga 2010, 2015.
5 While I was doing my fieldwork for the longer project (see Salonga 2010), I also participated in two international call centre communication conferences in Manila, the first and second "Talking across the World" conferences held in February 2006 and May 2007, respectively. Call centre visits were part of these conferences.
6 The first time that an informant is introduced, his or her position in the call centre industry is mentioned. All names used are pseudonyms.
7 In cases where the informants use Filipino or Tagalog or codeswitch between English and Filipino or Tagalog in the interview extracts, the English translations follow in parentheses.

8 For my informants, the term *rednecks* refers here to a particular group of Americans whose accent is difficult to understand compared to that of other groups of Americans.

9 These characteristics may also be coded as feminine and seen as a part of the feminization of talk that is taking place in the call centre industry. I explored this idea in the original study from which the interview data are taken (see Salonga, 2010). I would also like to note that the slogan of the Contact Center Association of the Philippines is "Serving the world."

REFERENCES

Aneesh, A. (2006). *Virtual migration: The programming of globalization*. Durham & London: Duke University Press. http://dx.doi.org/10.1215/9780822387534

Bayani, A.B. (2014). Bagong Bayani Awards. Retrieved 15 June 2014 from http://www.bbfi.com.ph/bagong-bayani-awards.

Billig, M. (1995). *Banal nationalism*. London: Sage Publications.

Blommaert, J. (2003). Commentary: A sociolinguistics of globalization. *Journal of Sociolinguistics*, 7(4): 607–23.

Blommaert, J. (2005). *Discourse: A critical introduction*. Cambridge & New York: Cambridge University Press. http://dx.doi.org/10.1017/CBO9780511610295

Blommaert, J. (2010). *The sociolinguistics of globalization*. Cambridge, New York: Cambridge University Press. http://dx.doi.org/10.1017/CBO9780511845307

Bolton, K. (2010). Thank you for calling: Asian Englishes and "native-like" performance in Asian call centres. In A. Kirkpatrick (Ed.), *The Routledge handbook of world Englishes* (pp. 550–64). London: Routledge.

Business Processing Association of the Philippines. (2008). Offshoring and outsourcing Philippines: roadmap 2010. Retrieved 15 June 2014 from http://www.bpap.org/bpap/index.asp?welcome2.

Cameron, D. (2000a). *Good to talk? Living and working in a communication culture*. London: Sage Publications.

Cameron, D. (2000b). Styling the worker: Gender and the commodification of language in the globalized service economy. *Journal of Sociolinguistics*, 4(3), 323–47. http://dx.doi.org/10.1111/1467-9481.00119

Cheng, S. (2010). *On the move for love: Migrant entertainers and the U.S. military in South Korea*. Philadelphia: University of Pennsylvania Press. http://dx.doi.org/10.9783/9780812206920

Contact Center Association of the Philippines. (2013). The Philippines. Retrieved 15 June 2014 from http://www.ccap.ph/index.php/the-philippines.

Convergys. (2012). *Pedestal*. YouTube video, one minute. Retrieved 15 June 2014 from http://www.youtube.com/watch?v=85mjjIW5uFw.

Convergys. (2013a). *Heroes Within*. YouTube video, eight minutes. Retrieved 15 June 2014 from http://www.youtube.com/watch?v=JDifkK_osw8.

Convergys. (2013b). *A Legacy of Heroes*. YouTube video, five minutes. Retrieved 15 June 2014 from http://www.youtube.com/watch?v=Cq8uIiLL3Ko.

Cowie, C. (2007). The accents of outsourcing: The meaning of "neutral" in the Indian call centre industry. *World Englishes*, 26(3), 316–30. http://dx.doi.org/10.1111/j.1467-971X.2007.00511.x

De Jesus, J.L. (2014). Call center agents vow to boycott GMA's "Borrowed Wife." *Inquirer.net*. Retrieved 17 June 2014 from http://entertainment.inquirer.net/131409/call-center-agents-vow-to-boycott-gmas-borrowed-wife.

Delias, D.M. (2014). Non-mobile labor and transnational subjectivity: Call center workers in global call centers in the Philippines. (Unpublished doctoral dissertation). National University of Singapore.

Department of Tourism. (2009). People and religion. Retrieved 15 June 2014 from http://www.tourism.gov.ph/SitePages/People.aspx.

Forey, G., & Lockwood, J. (2007). *"I'd love to put someone in jail for this"*: An initial investigation of English in the business process outsourcing (BPO) industry. *English for Specific Purposes*, 26(3), 308–26. http://dx.doi.org/10.1016/j.esp.2006.09.005

Freeman, C. (2001). Is local: global as feminine: masculine? Rethinking the gender of globalization. *Globalization and Gender*, 26(4), 1007–37. http://dx.doi.org/10.1086/495646

Friginal, E. (2007). Outsourced call centers and English in the Philippines. *World Englishes*, 26(3), 331–45. http://dx.doi.org/10.1111/j.1467-971X.2007.00512.x

Guevarra, A.R. (2010). *Marketing dreams, manufacturing heroes: The transnational labor brokering of Filipino workers*. New Brunswick, NJ: Rutgers University Press.

Heller, M. (2011). *Paths to post-nationalism: A critical ethnography of language and identity*. Oxford, UK: Oxford University Press. http://dx.doi.org/10.1093/acprof:oso/9780199746866.001.0001

Lockwood, J., Price, H., & Forey, G. (2008). English in Philippine call centers and BPO operations: Issues, opportunities and research. In M.L.S. Bautista and K. Bolton (Eds.), *Philippine English: Linguistic and literary perspectives* (pp. 219–41). Hong Kong: Hong Kong University Press.

Lomibao, M.A.L.L. (2007). Hanging up: A postcolonial analysis of offshored call center agents based on the narratives of female call center agents. (Unpublished master's thesis). University of the Philippines, Diliman.

Lorente, B. (2007). *Mapping English linguistic capital: The case of Filipino domestic workers in Singapore*. (Doctoral dissertation). National University of Singapore.

Lorente, B.P. (2012). The making of "workers of the world": Language and the labor brokerage state. In A. Duchene & M. Heller (Eds.), *Language in late capitalism: Pride and profit* (pp. 183–206). New York: Routledge.

Martinez, C. (2006). Welcome to Intelstar. In *Laugh trip: Dalawang Komedya* (pp. 1–20). Quezon City, Philippines: Milflores Publishing.

Milroy, L. (1980). *Language and social networks*. Oxford: Basil Blackwell.

Mirchandani, K. (2004). Webs of resistance in transnational call centres: Strategic agents, service providers and customers. In R. Thomas, A.J. Mills, & J.H. Mills (Eds.), *Identity politics at work: Resisting gender, gendering resistance* (pp. 179–95). London: Routledge. http://dx.doi.org/10.4324/9780203358269_chapter_10

Mirchandani, K. (2008). The call center: Enactment of class and nationality in transnational call centers. In S. Fineman (Ed.), *The emotional organization: Passions and power* (pp. 88–101). Oxford: Blackwell Publishing.

Mirchandani, K. (2012). *Phone clones: Authenticity work in the transnational service economy*. New York: Cornell University Press.

Mirchandani, K., & Maitra, S. (2007). Learning imperialism through training in transnational call centres. In L. Farrell & T. Fenwick (Eds.), *Educating the global workforce: Knowledge, knowledge work, and knowledge workers* (pp. 154–64). London: Routledge.

O'Malley, R. (2013). *Philippines Call Centre Outsourcing*. Accessed June 15, 2014. http://www.call-centres.com/philippines_call_centre_book.pdf.

Pennycook, A. (2001). *Critical applied linguistics: A critical introduction*. Mahwah, NJ, and London: Lawrence Erlbaum Associates.

Philippines: The new outsourcing hub. (2011). Philippines: Manabat Sanagustin and Co., Business Processing Association of the Philippines, and Team Asia. Retrieved 23 July 2013 from http://www.bpap.org/publications/research/phil-new-outsourcing-hub.

Poster, W. (2007). Who's on the Line? Indian call center agents pose as Americans for U.S.-outsourced firms. *Industrial Relations, 46*(2), 271–304. http://dx.doi.org/10.1111/j.1468-232X.2007.00468.x

Rafael, V. (1997). Ugly Balikbayans and heroic OCWs. In "Your grief is our gossip": Overseas Filipinos and other spectral presences, *Public Culture, 9*(2). Retrieved 17 June 2014 from http://www.bibingka.com/phg/balikbayan/.

Rodriguez, R. (2002). Migrant heroes: Nationalism, citizenship and the politics of Filipino migrant labor. *Citizenship Studies, 6*(3), 341–56. http://dx.doi.org/10.1080/1362102022000011658

Roxas-Tope, L.R. (1998). *(Un)framing Southeast Asia: Nationalism and the postcolonial text in English in Singapore, Malaysia and the Philippines*. Quezon City, Philippines: University of the Philippines Office of Research Coordination.

Rupert, M., & Solomon, M.S. (2006). *Globalization and international political economy: The politics of alternative futures*. Oxford: Rowman & Littlefield.

Salonga, A. (2010). Language and situated agency: An exploration of the dominant linguistic and communication practices in the Philippine offshore call centers. (Unpublished doctoral dissertation). National University of Singapore.

Salonga, A. (2015). Performing gayness and English in an offshore call center industry. In R. Tupas (Ed.), *Unequal Englishes: The politics of Englishes today* (pp. 130–42). Basingstoke, UK: Palgrave Macmillan.

Shome, R. (2006). Thinking through the diaspora: Call centers, India, and a new politics of hybridity. *International Journal of Cultural Studies, 9*(1), 105–24. http://dx.doi.org/10.1177/1367877906061167

Sonntag, S. (2009). Linguistic globalization and the call center industry: Imperialism, hegemony or cosmopolitanism? *Language Policy, 8*(1), 5–25. http://dx.doi.org/10.1007/s10993-008-9112-9

Srivastava, S., & Theodore, N. (2006). Offshoring call centers: The view from Wall Street. In J. Burgess & J. Connell (Eds.), *Developments in the call centre industry: Analysis, changes and challenges* (pp. 19–35). London: Routledge.

Taylor, P., & Bain, P. (2005). "India calling to the far away towns": The call centre labour process and globalization. *Work, Employment and Society, 19*(2), 261–82. http://dx.doi.org/10.1177/0950017005053170

Taylor, P., & Bain, P. (2006). Work organization in Indian call centres. In J. Burgess & J. Connell (Eds.), *Developments in the call centre industry: Analysis, changes and challenges* (pp. 36–57). London: Routledge.

Tiatco, S.A. (2014). Theatre, entrapment, and globalization in welcome to Intelstar. *Humanities Diliman Review, 11*(1), 1–28.

Tupas, T.R. (2001). Linguistic imperialism in the Philippines: Reflections of an English Language teacher of Filipino overseas workers. *Asia-Pacific Education Researcher, 10*(1), 1–40.

Tupas, T.R. (2004). The politics of Philippine English: Neocolonialism, global politics, and the problem of postcolonialism. *World Englishes, 23*(1), 47–58. http://dx.doi.org/10.1111/j.1467-971X.2004.00334.x

Tupas, T.R. (2007). "Go back to class": The medium of instruction debate in the Philippines. In L.H. Guan & L. Suryadinata (Eds.), *Language, nation and development in Southeast Asia roundtable* (pp. 17–38). Singapore: Institute of Southeast Asian Studies.

Tupas, R., & Salonga, A. (2016). Unequal Englishes in the Philippines. *Journal of Sociolinguistics, 20*(3), 367–81. onlinelibrary.wiley.com/doi/10.1111/josl.12185/full

Vaish, V. (2008). *Biliteracy and globalization*. Clevedon, UK: Multilingual Matters.. http://dx.doi.org/10.1007/978-0-387-30424-3_40

6 Transnational "Homies" and the Urban Middle Class: Enactments of Class, Nation, and Modernity in Guatemalan Call Centres

LUIS PEDRO MEOÑO ARTIGA

Introduction

Since its emergence the global call centre industry has had considerable economic, social, and political impact on Guatemala City. Large advertisements for call centre companies dominate the employment section of the national newspapers. Many call centres are located in large shopping centres that have a commanding presence in Guatemala City and where it is now common to find groups of workers speaking English during their breaks. In the one public university and the numerous private universities of the city, call centre employment is now a typical way to pay for tuition. Over the past ten years employees of transnational call centres have become a new and distinct urban sector, composed of more than twenty thousand workers (Maul & Bolaños Fletes, 2013, p. 7). Most are young, middle-class university students with monthly incomes that place them at the level of highly skilled professionals. Employees have their own cars and make conspicuous use of shopping malls. These malls often act as islands where middle-class workers can live and work in relative comfort and security in a highly segregated, impoverished, and violent city.

The global call centre industry has adapted itself to the Guatemalan neoliberal order. It offers employment to urban middle-class youth, promoting consumerism, while attaching to and reinforcing the urban spatial model that is dominated by privately secured spaces of socialization and consumption. It is presented as a form of qualified work for the middle-class section of society, which has traditionally embodied the image of modernity in Guatemala. The industry generates public discourses that promote the image of a new young, cosmopolitan worker who at the same time is proud to be Guatemalan. This worker, thanks to the talent and passion for the work, is capable of achieving personal success in the global service market. The government promotes the call centre industry, emphasizing its high capacity for generating "qualified" and well-paid employment. While making alliances with the public sector (O'Neill, 2012, p.21), call centres project a powerful image of modernity. Organizational representatives advocate for higher levels of flexibility in labour legislation and for reforms to foreign investments laws (Meoño Artiga, 2011, p.15). In general, both the state and the business sector present the call centre industry as an opportunity for Guatemala to insert itself into the globalized knowledge-based economy.

However, in Guatemala, call centres are also associated negatively with the sweatshop industry. In conversations with people involved in this sector the word *maquila* (a common term in Latin America for "sweatshop") almost always emerges. My own interviews with workers often pivoted upon the connection between call centres and *maquilas* as an expression of labour exploitation (Meoño Artiga, 2011). Chief executives and managers in public interviews tend to distance themselves and call centres from the *maquila* manufacturing industry.[1] In addition, call centres are often denounced, both in public and in private discourses, as dangerous or indecent spaces.[2]

The disputes and contested representations about the character of the new global workplace in the city are fuelled by the incorporation of a new social group into this industry, one that is not identifiable by the traditional characteristics of the Guatemalan middle class. It is a group composed of a growing minority of English-speaking workers who have been deported from the United States and find in call centres almost their only option to access the formal Guatemalan labour market. These returned migrants have become a fundamental labour force for the call centre industry. There is no official data regarding the number of deportees or returned people working in Guatemalan call centres; however, Kevin L. O'Neill (2012, p. 5), who conducted ethnographic research inside the biggest call centre in the country, says that this population has no doubt been integral to the industry: "Guatemala's current surge outpaces its middle-class, bilingual population. By 2012, experts predict, Guatemala's call center industry will reach full saturation ... This is also why a range of stake holders, from security officials to CEOs, from clergy to councilmen, now make a rather desperate pitch to a rather desperate population. Call centers *need* the deported."

In my own fieldwork, informants reported that some 10 to 30 per cent of the total call centre workforce is made up of deported or returned employees. Recently several middle-sized companies have emerged in which deportees or migrants comprise the *majority* of workers. The official website of World Connection, a mid-size Guatemalan call centre, promotes the "U.S. cultural understanding" of its workforce, stating, "Most of our agents have lived or spent time in the United States."[3] This characteristic is confirmed by my own field interviews.

The presence and growing importance of deportees in global call centres has been documented in a small but growing number of academic and media reports – in Mexico (J. Anderson, 2013; Castillo, 2012; Da Cruz, 2012; *La Jornada*, 2014; Romero Loyola, 2012), Guatemala (Escalón, 2011; Hurtado, 2010; Meoño Artiga, 2011; O'Neill, 2012; Palencia Frener, Mendizábal Juárez, & Poroj Abaj, 2011), and El Salvador (Da Cruz, 2012; Rivas, 2007).[4]

Central American discourse on deported men and women inside the call centre industry focuses on the negative characteristics of this group, portraying deportees as dangerous criminals or gang members (Meoño Artiga, 2011, p. 94; Rivas, 2007, p. 30). In the Guatemala call centre world these workers are sometimes known as *homies*, a stereotypical term for deported migrants who are assumed to be gang members. The significance of the term *homie* helps to explain the tensions caused by the inclusion of this segregated urban labour force. *Homie* is a term from U.S. urban culture that is symbolically linked to black and Latino gangs. In Guatemala it is used to describe all those who have returned from the United States, even if they have had no previous affiliation with a gang. The term also describes the body aesthetics displayed by deported and other employees in the workplace, which may include the use of tattoos, baggy pants, and loose shirts, defined as typical of southern California. The relationship between homies and other call centre workers is marked by tension. Some of this tension can be explained by fear and profound mistrust, and some of it can be understood as an allure that comes with travel. It can be said that homies working in call centres represent the new urban pariah – a racialized group of workers socially constructed as gangster and inherently dangerous, owing to their social origin, life trajectory of migration, and deportation history. Nevertheless, the call centre sector is also successfully placed in the "strategic point where the borders of class, race, work and nation intersect" (Mirchandani, 2012, pp. 23–6).

In this chapter I analyse the tensions of the global and flexible call centre industry as it integrates a diverse pool of workers in its productive process. Faced by a local logic of hierarchical segmentation and urban segregation, corporations struggle to construct a more socially homogeneous labour force. I will argue that the merger of two differentiated segments of the population in the Guatemalan call centre industry – deportees on the one hand, and middle-class university students on the other – is uncomfortable and questioned. The ongoing establishment of this new transnational call centre sector disrupts a traditional order that clearly establishes the place of each class and racialized group in the formal Guatemalan labour market. The presence of homies within call centres that are strongly associated with the middle class challenges the rigid model of class stratification and urban segregation in Guatemala City. This is especially apparent if we take into account wage rates. In Guatemala the monthly minimum wage is equivalent to US$273.54. Yet up to 86 per cent of the population earn less than this figure. The average income in the public sector is US$189, and in the private sector US$240 (Maul & Bolaños Fletes, 2013, p.43). In contrast, a bilingual telephonic agent has an average salary of US$450, which puts his or her earnings well above most other working and middle-class groups. Therefore, the call centre subverts the traditional correlations between education level, social origin, and life trajectory in terms of reaching a "professional" level of income. Consequently the call centre emerges as an ambiguous and ruptured space, in which the characteristics of belonging to the middle class are both re-enacted and disputed.

The tensions that I highlight, between global liberal trends of diversity management (Poster, 2008, 2007a, 2007b) and local (and more rigid) logics of exclusion, have had important consequences for labour in terms of separating and differentiating groups of workers within the industry. It has also produced differentiated business policies as companies try to deal with different imaginaries of the workers who are supposed to comprise this modern labour force. In general the conflicts represent a symbolic dispute over the characteristics of the new globalized Guatemalan worker.

This chapter is organized into three sections. The first delineates the main characteristics of the global call centre industry, and the particularities of its insertion into the urban context of Guatemala City, emphasizing the logics and practices of segregation and discrimination on an urban level. The second describes the functioning of these logics in the national formal labour market (which I will later contrast to the global

call centre labour market). The final section focuses on the transformations of traditional models of exclusion, new enactments of class and
nation, and the persistence of the practices of stigmatization and fear
that shape worker's relations.

The analysis in this chapter is based on eleven months of ethnographic
fieldwork in Guatemalan transnational call centres. I conducted the
first phase during the last three months of 2010, and the second phase
during the first seven months of 2013. After initially failing to obtain
enough reliable information through traditional methods, I enrolled as
an agent. I worked for just over two months in two of the city's major
call centres.[5] Two of the main investigative tools used during this period
were participatory observation and a detailed field diary. The second
phase of the investigation involved in-depth interviews with employees
of several call centres, as well as English trainers, government officials,
and academics related to the industry. In total, I conducted ten in-depth
interviews and thirty-eight shorter, structured interviews with workers
of seven call centres.

The Global Call Centre Industry and the City

Guatemala City

Guatemala has one of the one of the highest homicide rates in Latin
America, and the highest rates of violence in the country's capital. The
daily crime that ravages the city has been traditionally considered to be
rooted in poverty and large income disparities, an abundance of weap-
ons, a legacy of state repressive terror and disrespect for human rights,
accompanied by weak law enforcement and judicial systems. As a result,
the city is characterized as being poor, largely uninhabitable, and de-
industrialized and having a deteriorating infrastructure. Guatemala City
is sustained by an economy that relies extensively on family remittances
and both informal and illegal commerce (Levenson, 2011).

Violence is concentrated in particular sectors of the economy although
its impact can be felt throughout the city. The two most dangerous urban
occupations are those of shopkeeper and public transport driver;[6] thus,
violence affects Guatemalans' daily life in two basic ways – popular
trade and public transportation (Dudley, 2011). Urban violence is
related to an extortion economy, associated with gang activities, that
victimizes the poorest inhabitants of the city. Since the signing of the
peace agreements in 1996 that ended the thirty-six-year-long armed

conflict subsequent governments have adopted the same tactics to deal with the increasing urban crime, specifically the problem of youth gangs. Policies developed around controversial "strong-hand" (Arana, 2005) laws that "advocate strong law enforcement approaches towards gangs or maras and involve questionable methods such as the arbitrary detention of suspected gang youth for simply wearing baggy pants or sporting tattoos" (Donaldson, 2012, p. 2)[7]. Popular support for these policies is explained in part by the consistent government and media discourses that have construed gangs, and popular youth in general, as the new enemies of society. Allegedly these social pariahs need to be suppressed to protect the nation, but in actuality the discourse defends the status quo (Levenson, 2013; Martel, 2006).

Guatemala's capital can be described as a city of walls. This concept, coined by Teresa Caldeira (2007), describes the growing patterns of urban segregation that are linked to rising urban violence and separate the upper- and middle-class sectors through the construction of fortified enclaves. According to Caldeira, the construction of physical and symbolic walls, and the consequent withdrawal from public life of the upper and middle classes, has broad repercussions with respect to citizenship and urban practices. The disappearance of public space and its substitution by fortified socially homogeneous enclaves promotes the territorial stigmatization of large segments of the population (Wacquant, 2009). This hinders interaction between diverse social groups and in general defines inter-class relations by fear and suspicion. Ideas of mixture and social heterogeneity associated with the openness of the city are increasingly perceived as dangerous. As a result, geographic isolation and private security measures become a subject of desire and status (Caldeira, 1996, p. 304).

Caldeira's model has been modified in order to account for the scale differences between a city like São Paulo, on which she based her analysis, and the middle-size cities of Central America. Rodgers (2004) and O'Neill and Thomas (2011) hypothesize that "the relatively small size of the urban elites" makes it difficult to establish autonomous gated communities in the region. They propose the concept of a "disembedded city" in which a whole layer of the city is detached from the urban grid through the constitution of a fortified network of small enclaves for the exclusive use of the urban elites (Rodgers, 2004). This nuanced analysis focuses more on infrastructure and stresses the strategic role of both private security and improved private transportation for the reorganization of the urban order in Central America. In this model the large shopping malls emerge as strategic nodes in the fortified network, used not only for consumption and socialization practices but also for productive processes, therefore facilitating the integration of the city into the global service economy.

The Guatemalan Global Call Centre Industry

The city saw the beginnings and accelerated development of the call centre industry in 2004. Since that time it has been one of the urban industries with the highest rates of growth. As stated above, there are currently more than twenty thousand people working in global centres throughout the city. The labour market is controlled by large multinational corporations that employ more than 70 per cent of the workforce. At the forefront are two large companies with more than five thousand workers each: the formerly Guatemalan-controlled Transactel, which was bought in 2014 by the Canadian-based Telus International Corporation; and Allied Contact Center (ACC), a BPO services provider founded in Guatemala, which is now based in the United States. A second grouping of companies employs more than two thousand workers and consists of Atento, formerly a Spanish multinational that was recently bought by the U.S.-based Bain Capital investment firm, and Expert Global Solutions, Inc. (EGS, formerly NCO), an American multinational based in Texas. The rest of the multinationals employ approximately one thousand workers. These include Genpact, originally from India and now domiciled in Bermuda with executive headquarters in New York City; 24/7 Customer, which also formed in India but is now based in California; CapGemini, a French multinational corporation headquartered in Paris; and ACS-BPS, a subsidiary of the U.S.-based Xerox Corporation. The remaining 30 per cent of the workers are employed by medium-size companies with between five hundred and a thousand workers each. Two of them are the the U.S.-based firms Serco Global Services and C3/Customer Contact Channels. The last one in this group is the previously Guatemalan-owned Innovative Contact Solutions (ICS), which was purchased in 2013 by the Canadian outsourcing firm 24–7 Intouch. Finally there is a group of smaller Guatemalan-owned companies with between one hundred and five hundred workers. These include Nearsol, World Connection, GuateCall, and 5ND.

Ninety per cent of the customers served in the Guatemalan industry are located in the United States, and most call centres offer bilingual

(English and Spanish) services. Some also offer services in other European languages, but these account for less than 4 per cent of the market according to a recent research report (Everest Global, 2014). The number of accounts served varies and fluctuates; however most involve sales and basic customer-support operations. CapGemini and Genpact are notable for specializing in finance, accounting, and back-office services. In general, big multinationals like Telus International and ACC have a diversified client portfolio serving several dozen accounts each. The size of the team handling an account can vary from a few workers to several hundreds. Many of these multinational companies are strongly associated with their major accounts, such as CapGemini with Coca-Cola, EGS with T-Mobile, ACC with Tracfone, Telus International with Telus, ACS-BPS with Xerox, and Genpact with GeMoney. Many of these big accounts have been relatively stable, staying with the same Guatemalan call centre for several years. For example, TXU Energy customer service and revenue management accounts have been serviced by Telus International in Guatemala City since 2005, and T-Mobile phone activations and customer support have been serviced by EGS since 2009.

In general, the smaller Guatemalan companies specialize in sales and telemarketing. They tend to serve much smaller American and Canadian enterprises. Some examples include Simply Platform Beds and Connect Your Home, whose sales processes are attended by Guate-Call and World Connection, respectively. Guatemalan small companies also seem to have greater levels of flexibility. They often offer shorter contracts, sometimes for high sales seasons only, or briefly incorporate large numbers of workers, many from other companies, for short-term events known as "Ramps."

The City and the Industry

The emergence of the call centre industry is related as much to the growing bilingual English-Spanish service market in the United States as it is to the availability of a workforce capable of attending to it. The principal generators of this new labour market include international migration[8] and a privatized and exclusive education system that favours English-language training for middle-class youth (Álvarez Aragón, 2010).

In fact call centre work is strongly associated with the urban middle class. Call centres are meeting places for thousands of university students, many of whom are working to fund their education. Also strongly represented are students who have recently graduated from middle- and upper-class private schools. These groups form the major-ity of the labour force. Housewives, unemployed professionals, and occasionally foreigners are also included in this form of flexible work.

However, second in importance in the call centre industry are the returned migrants, many of whom have come back to Guatemala through deportation from the United States. These workers are known as *homies* or *cholos*, and their Latin street style is strongly associated with a criminalized and exiled identity. Homies (and others who dis-play this style in the workplace) are seen as young, transnational gang members, or *mareros*, who are blamed for causing the high level of crime in Guatemala.

Nevertheless, as my field research revealed, call centres are one of the only formal workplaces in which this youthful style may be relatively accepted. The "liberal corporations" encourage workers to "just be themselves," promoting the expression of unique individuality and tolerating different lifestyles within the workplace (Fleming, 2009). Call centres embrace a management system that, according to Peter Fleming and Andrew Sturdy (2011), seeks to capture "sociality, energy and 'authentic' or 'non-work' personalities as emotional labour." This creates a paradox in an environment that is otherwise presented as safe and decent.

Call centres are spaces, as nowhere else in the city, where interac-tion is possible between two urban segments of the population that are traditionally segregated, both spatially and socially. Therefore, the emergence of homies in the call centre industry is an anomaly within a rigid social order. According to this logic, members of the Guatemalan middle class should be the most appropriate group to lead Guatemala through its insertion into the modern globalized labour market. Yet, interaction between these social groups distorts the functioning of the exclusion narratives that characterize Guatemala. A colonial mentality of servitude has traditionally used racist and authoritarian practices to exclude large segments of society. Such a mentality focuses especially on those who have been traditionally stigmatized as dangerous and who represent the greatest obstacle to development in Guatemala: the indigenous and mestizo urban poor. These peoples are placed in opposition to the highly educated (and socially constructed as more "white") middle class.

As a result the symbolic construction of transnational call centre work as truly "professional" is problematic and contested. In fact, the notion of professionalism itself becomes an area of dispute. Some associate call centre work with professionalism owing to the relatively high class and educational level of its workers, while others, adopting a more neoliberal discourse, propose a new notion of professionalism, associating the ability to speak English with "talent," and defining adherence to the hyper-flexible work ethic as "passion."

In the next section I examine some of the exclusionary logics and practices of segregation in the Guatemalan labour market, to later contrast them with those in transnational call centres.

Employability and Stigma in National Programs

In this section I argue that the authoritarian and racialized logics of exclusion that have permeated social life in Guatemala City have entered the formal urban labour market. Access to formal employment for large social groups (especially for youth from the poor zones of the city and for deported migrants) is obstructed through discriminatory and criminalizing practices. These practices have been normalized within the Guatemalan labour market through a racialized discourse of "security." The discourse has been widely accepted by all members of the economy, from business owners to employees, and even by government officials and social activists who supposedly promote more inclusive employment policies.

TERRITORIAL STIGMATIZATION

In Guatemala City a large majority of youth has been excluded from formal employment through a process of "territorial stigmatization" (Wacquant, 2008; 2009), which superimposes exclusions based on race and class onto labour. This stigmatization criminalizes poor urban youth who are residents of areas that are considered to be vulnerable, by denying them access to work on the grounds that they are dangerous. Youth are constructed as potentially *mareros* (slang for members of an organized gang) and *choleros* (slang for low-class individuals, especially of mixed or indigenous heritage), thus as dangerous and without culture at the same time.

This is illustrated through the Youth Employability project, implemented in Guatemala in 2013. The project was a private initiative that tried to introduce three hundred youth, residents of "vulnerable urban areas," to formal employment in national companies. The program was carried out by the non-governmental organization Alianza Joven, which maintains strong ties to Rotary clubs in Guatemala and, through them, to the business sector. The United States Agency for International Development (USAID) financially supported the initiative. The Youth Employability project used an internship model, in which youth received a monthly stipend during training from Alianza Joven, and then on-the-job training for two months from a company. Those responsible for the project claimed that financial support during training would facilitate employability afterwards (interview, 2013). The project was considered to be very successful, and the model has become the basis for the current National Employment Policy. The government of Guatemala expects to create forty thousand new formal positions using the same outline of training and paid internships.[9]

Yet, the experiences of rejection, discrimination, and criminalization by many participants of the project expose the racialized mechanisms of exclusion by Guatemalan companies. The most generalized of these is the refusal to hire residents from areas of the city considered "red zones." This was expressed by Liliana,[10] who was responsible for the project: "If you're from Zone 18, [the companies say] they won't hire you, period. It's a question of security" (interview, 2013). Zone 18 is the most populated zone in the city, with more than 225,000 inhabitants. It also is the zone with the highest murder rate.

The practice of territorial stigmatization is illegal but used extensively. It is accompanied by a series of security procedures, which exclude job applicants whose characteristics are considered suspicious. Physical examination of tattoos on aspiring workers is one of these practices, justified by employers as the need to identify whether the individual is a member of a gang. One of the most problematic practices, also used extensively, is the use of a polygraph test. According to those in charge of the Youth Employability project, the majority of participants took such a test, and in fact it was a requirement for all applicants in service positions that required handling cash (in other words, the majority of positions available in the Guatemalan urban service economy). Many of the participants were rejected in the first round of polygraph tests. However, through pressure by Alianza Joven, employers reconsidered the results or allowed applicants to repeat the polygraph. At that point, Alianza Joven trained students to respond better to interrogators. For example, the youth were told to "try not to lie" and to "answer 'yes,'" if asked about knowing anyone involved in criminal activity (interview, 2013). Despite these issues, managers of the youth employment project were positive about the outcomes. According to the program director, one of their major achievements was enabling youth to overcome barriers of exclusion. They attributed this to the legitimacy of the Rotary business owners involved in the project, as well as the financial support by USAID of their training (interview, 2013).

However, suspicion and stigmatization did not end with the hiring of the youth involved in the project. Many of the participants encountered problems later on, such as situations in which their bosses and co-workers were afraid to work with them. Fear was sparked for various reasons: in some cases, when new employees told stories of violence occurring in their neighbourhoods; in other cases, when the youth exhibited "suspicious conduct," like showing "too much interest in the boss's car" (interview, 2013).

As stated above, these practices are not typically considered discriminatory by employers. Rather, they are seen as security measures that are undertaken by the owners of the companies and, in some cases, demanded by the other employees. Those in charge of the employment project, who believed they were helping youth overcome barriers of exclusion, did not recognize such actions as major obstacles to employment. Instead, project coordinators claimed that these problems occurred for several reasons: the "bad attitude of youth towards work – they seemed to want to start as the boss"; the educational deficiency of the applicants and, in turn, their lack of technical capabilities; the violence of vulnerable areas and its negative effects on residents; and finally, a "paranoid attitude" on the part of the business owners as well as the workers in the company (interview, 2013).

DEPORTATION STIGMATIZATION

Since the late 1990s the city of Guatemala has received a growing number of people deported from the United States. The year 2013 marked a new record, when the number of migrants deported to Guatemala exceeded fifty thousand (*La Hora*, 28 December 2013). The deportees share many of the same socio-economic, cultural, and spatial characteristics as those of the residents of the vulnerable areas of Guatemala City. This growing segment of the population, considered to be an urban pariah, like the poor urban youth, is stigmatized and excluded from formal employment (migrants who are fluid in English are the exception because they are eligible for, and thus included in, the call centre industry workforce).

In the specific case of Guatemalan call centres the ability to speak English allows returned migrants to surpass barriers of exclusion and insert themselves into formal employment that is considered decent work.

It is identified as such not only because of the high level of income and education among the workers but also because of the place in which the work is carried out, namely huge corporate offices, located in large malls throughout the city, especially in the central commercial district.

However, as I will argue, the insertion process for deported migrants is not as inclusive as it might seem. Hiring policies fluctuate within the industry, from full acceptance to total rejection of the deportees. Furthermore, inclusion produces racialized conflicts[11] and disputes. These disputes extend through the production process and affect the relations inside each company as the workers are marked by fear.

The continued stigmatization of deportees was articulated by Congressman Jean Paul Briere,[12] the president of the governmental initiative "Welcome Home," a public-private alliance that attempts to increase the inclusion of deportees into the labour market. This initiative predictably faced many of the same problems as did the Youth Employability project. In fact, some of the problems were even more acute, as prospective employees carry the stigma of deportation *along with* that of criminal activity. To counter this stigma, Congressman Briere called for an abandonment of the use of the word *deportado* (deported). In a televised interview he pointed out "that many [returnees] come [back] with technical abilities acquired after many years there that they could apply here."[13] As an example, Transactel, one of the major call centres in Guatemala associated with the "Welcome Home" initiative, selects migrants with "many years there" for their workforce.[14] There is a clear demand for bilingual deportees. As a result, transnational call centre companies participate in the "Welcome Home" program, as well as other employment policies for deported migrants. Yet, the process of constructing the new globalized Guatemalan worker generates multiple conflicts.

Paradoxes of the Global Call Centre: Between Inclusion and Exclusion

The labour force in Guatemala's global call centres stands out for being more diverse than that in national service companies. Along with the mostly young university students, returned migrants (mostly deported), housewives, and unemployed professionals make up an important percentage of the workers. To enter into this workforce the universal prerequisites are fluency in English and the ability to operate a computerized database with efficiency – two abilities acquired in very different forms by multiple segments of the labour force. Although the call centre industry appears to challenge many of the conventions of labour exclusion, it maintains many of the same stigmatizing practices and discourses. These can be seen as policies that in some ways favour, but also hinder, the massive incorporation of bilingual workers into the global service market. I will now examine this dual tension in three aspects of the global call centre industry: recruitment, advertising, and hiring; social and occupational segregation on the shop floor; and stigmatization of deportees as criminals.

RECRUITMENT

The process of recruitment in the call centre industry, unlike the national industries discussed earlier, is flexible, open, and fast. Hiring in some companies is done on the same day that the candidate applies, taking an average of four hours (Meoño Artiga, 2011, p. 77). Indeed, employers are quite eager to gain new employees. In addition, they do not use criminalizing practices such as giving polygraph tests or physical examinations.

On the contrary, many of these companies use a liberal discourse in their publicity campaigns, celebrating the differences and alternative styles of youth groups – to the extent that even dressing in the cholo style can sometimes be considered appropriate for decent and professional work. The call centre industry, in this sense, is more open and successful in incorporating various social strata into its labour force than are the national companies. The global companies articulate a discourse of inclusion that frequently portrays housewives and youth as members and beneficiaries of this new flexible form of employment. In turn, this discourse allows the industry to demand more flexibility in national labour legislation through political lobbying.

In the process of incorporating the most diverse pool of workers, the industry also expands the representation of professional work in Guatemala. Call centre firms emphasize that employment is gained through "talent" rather than level of education. This is apparent in their hiring policies and media discourses. Take, for instance, Transactel, now a subsidiary of the Canadian corporation Telus Communications, which has the largest call centre in Guatemala. Its main office and corporate towers are part of the Galerías La Pradera mall and host more than two thousand employees. The call centre dominates the highway from Guatemala City to El Salvador. This peripheral mountain area is the most expensive and modern zone and has the greatest concentration of middle- and upper-class residents in the city. Transactel's publicity image includes a young "hero, proudly born and raised in Guatemala ...

[who] fights for his talent." The notion of professionalism is malleable to attract the returnees.

The promise is an upwardly mobile lifestyle. Guatemalan Transactel workers are the central characters of the company's advertising campaigns, which feature a wide range of actors: from an Afro-Guatemalan to a cholo, a university student to a hipster. All appear in the same photo session, either celebrating "Movember" (an international event during the month of November, involving the growing of a moustache to raise awareness for men's health), or dancing in a viral video accompanied by pop star Pharrell Williams's song "Happy." The company presents an image of a heterogeneous labour collective that appears to have sufficient income for living a "decent" and modern life. In Guatemala City this means having minimum daily contact with social groups considered dangerous. It also means owning a car, living in a gated community, and, especially, participating in the closed spaces of socialization and consumption, such as churches, universities, private sports clubs, and large shopping centres.

Yet the conflicts between inclusion and exclusion are evident when examining the differences in the companies' hiring policies, which can range from an almost complete acceptance of deportees to a total rejection. One example of the latter is IndiaCorp, an Indian-owned and -operated company in Guatemala. This company is strategically located adjacent to the largest, and only public, university in Guatemala. Its marketing discourse downplays the liberal, youthful, and fun character of the job and emphasizes the professional image of the company as a "wholesome environment" (Meoño Artiga, 2011, p. 59). In contrast to Transactel, IndiaCorp projects a much more unified and homogeneous labour picture in its advertising images and discourses, specifically, that of a young professional or university student who is urban and white and has a global lifestyle. During a training session in which I participated, Mafer, a young English teacher with more than five years of experience in the industry, explained how other types of workers are shunned: "Here at IndiaCorp, despite their non-discriminatory policies; they wouldn't contract you if you worked illegally in the United States, or even if your parents did. They also wouldn't contract you if you are in debt or pregnant."

Another example of the former case is GuateCorp, one of the smaller Guatemalan companies in the market that employs many deported workers. The austere publicity and marketing discourse of this company claims that they are simply looking for workers with a high level of English, which they define in numerical terms as 95 per cent proficiency. This may be a way to reference English fluidity outwardly, while hiding that in reality they are looking to hire deported migrants. This indirect way of depicting deported workers contrasts to the more direct forms of nationalized recruiting used in El Salvador. There I have found companies stating explicitly that they are looking to hire "natives" for certain accounts.

Although the call centre industry loosens patterns of labour exclusion in Guatemala, it does not eliminate them. Instead we see polarization between urban groups, and social and symbolic borders between workers. This is described by Christine, a young industrial engineer. At the time of the interview in 2013 she was unemployed but had worked for three years at SpainCorp, a call centre located on the upper level of a large mall: "I think there is a certain separation between the groups [deportees and students]. Like, each tries to make its own little groups, and in the end the people who came from the United States join with others that came from there too. There is always the separation of groups."

This social segregation is confirmed by Gustavo in his experience at UsaCorp, where he worked for nearly four years. Thirty years of age, Gustavo resides in a poor neighbourhood of Guatemala City. He learned English through a private scholarship to attend high school at the Colegio Americano, one of the most expensive and prestigious schools in Guatemala City, which is traditionally attended by children of foreigners and of the country's elite. UsaCorp is a transnational corporation that integrates a large number of deportees into its workforce. It is located in the Galerias Primma, another one of the big malls in the city, and employs a staff of more than one thousand employees. Yet, as Gustavo describes, social groups mark very clearly defined boundaries: "The homies have their groups apart, but we get along well. They are broken up in teams, but they only get together with each other. For example, there are two big [drug] dealers in the company – one that sells to the homies and one that sells to everyone else. It's like each one has their territory, and they don't mess with the other." Call centres have socially marked "territories," mimicking the urban spaces outside. These territories reflect the relations of gangs and drug politics in Guatemala City.

Discrimination is also present in the labour policies of the companies that segregate workers by campaign. Transactel divides its eight thousand employees into two campaigns: regular accounts (with large percentages of migrants) and elite accounts (without migrants). Wicho affirms this pattern. He is a thirty-year-old with a university education and has worked for the last five years in the information technology department of Transactel: "Look, there's everything; some accounts are shit (like Sodom and Gomorrah) – principally in sales, where there are more homies. But there are also really expensive accounts, like those for accounting, or the ones in German or French." Elite accounts, therefore, involve higher level finance abilities and multi-European language skills. Regular accounts deal largely in telemarketing, requiring lower skills, and are more unpleasant for workers. Through this example we see the influence of racialized stigmatization of labour inside the companies. Managers see homies as more appropriate for sales, which supposedly requires an "aggressive" style, rather than for customer care, which requires friendliness and servility.

The racialized stigmatization of labour is also especially common in a recent and growing sector of the call centre industry – medium-size companies, which have a higher share of deportees specializing in sales. This is the case for GuateCorp, with around three hundred employees. Deportees make up the only workforce for some of their accounts. Vico, a young Costa Rican student who is studying for a master's degree in social sciences at a private university and pays for his studies by working at GuateCorp, states: "It's brutal where I am. Many homies have been killed because they walk around thinking this is Los Angeles. But here [in Guatemala], they break them. [*Homies*] *only work in sales. There aren't any support accounts.* [For those,] they [managers] just want stupid kids who don't know the shit [that] you can do with sensitive client information [social security numbers and credit card numbers]. Plus, for sales, you need better English [to] understand ... everything and be able to instantly respond to whatever shit comes out of the client" (emphasis added).

Managers exclude migrants from customer support jobs because they lack the appropriate skills to take charge of clients, despite the fact that they have the neutral accents required. This is supposedly due to a lack of "culture," defined in general terms as the capacity of a Guatemalan to offer a friendly and helpful tone on the telephone with North American clients.

CRIMINAL STIGMATIZATION OF RETURNED MIGRANTS
Despite being actively sought and hired by the call centre industry, homies are frequently criminalized. The chief executive officer of Transactel, Guillermo Montano, a young businessman and a member of the

economic elite, described the deportees in these terms in an interview in 2010: "[They are] like a virus that gets in and complicates everything." He also referred to the inability of the company to distinguish between migrants and gang members and complained further about the difficulty to eradicate the characteristic American-Spanish accent from the returned migrants (Hurtado, 2010).

The archetypal discourse of criminalization is reproduced along the service production chain in more or less similar ways. Call centre managers suspect an employee who is a returned migrant of being addicted to drugs and involved with gangs. Returned migrants are characterized as violent and potentially dangerous to their co-workers. Mafer, the English trainer at IndiaCorp, stated: "The problem with Latino people who grew up in the USA is that they have a difficult attitude, of resentful homies. ... [They] have great English, honestly. Maybe not computer skills. But the way they relate to each other – man, that sucks. For example, I had Angel and Emmanuel in a training batch. They were cholos from gang A and gang B. Finally, they got into a fight inside the company, so they got fired. That's how the company made an investment in people who didn't deserve it."

Arguments about migrants as criminals are reinforced by other call centre industry personnel, like Donald Montana. He works for an independent recruitment company that finds employees for call centres, offering English courses and client services to those who aspire to enter this market. (Interestingly, he is an English teacher of Guatemalan parents, who was born and raised in the United States. He changed his name to one that he considered to be more "American" and engaged in a legal battle in Guatemala to change his identification papers to state that his place of birth was the United States.) Explaining the way in which the stigma of returned migrants affects the hiring policies of companies in the sector, he stated: "There are call centres that don't accept deportees simply because they are criminals, they haven't gone to university, they don't have a good level of English, they use a lot of slang and language from the street, but also for the aesthetic problem. On the other hand, they have an American accent, they use the mannerisms, and above all they have the American mentality. They are ready to 'play the long game.' They don't go around like Guatemalans, [avoiding work by] saying, 'It's just that I have to take care of my grandmother.' They say, 'Okay, let's do it.'"

The discourse of criminalization against homies extends even to the workers. In multiple interviews university students mentioned that the presence of returned migrants was one of their major factors in choosing a company or an account. This was true for Eliza, a graduate of a private university and a former employee of FrenchCorp, which is a French transnational and one of the most prestigious call centres in Guatemala. During an interview, she said: "What I like about the work, and why I lasted so long, is because of the atmosphere and the people there, because they're people at another level. They don't treat you badly or give you mean looks or look at you with resentment. Not like one of my friends who worked at Transactel. In a group of ten that she was in, maybe six were undocumented. [She told me] they were always using swear words and kept giving her bad looks, so she didn't last."

The deported label has meaning for *potential* employees too. Precisely because of the large presence of returnees at GuateCorp, the company appears far less attractive to many (non-returnee) workers – despite offering wages that are comparable or even superior to those of the large transnational corporations. On this basis of nationality, workers make distinctions between good and bad companies, as well as between good and bad accounts. These values refer, in many instances, to the absence or presence of migrants, as much as to the level of English, the technical training needed, and the amount of wages offered.

In general, workers do not perceive these observations and generalizations about homies as discriminatory but as legitimate discourses of security. This is the case of Christine, the young industrial engineer quoted above. When talking about student-homie relations at Spain-Corp she stated:

> The thing is, for people who come from college, like me, maybe they're scared of being around this type of people. Although I never [had] any problems, really [...] My boyfriend told me that right now [at UsaCorp2] there are some there, he called them homies, that [wear] T-shirts under their shirts, with their shirts buttoned at the top and the rest open.
>
> So, he said that one day this guy came up to him, saying, "You don't know what I've been through!" Because they're like that – really angry. So then my boyfriend said, "But calm down!" But the guy kept freaking out and was almost pushing [my boyfriend], trying to create conflict inside. Some [homies] fight outside the call centre, and they had to fire them.
>
> Another problem is that they bring drugs inside the call centre ... The person who sold wasn't a homie; he dressed like them, but he had never

been to the United States or anything. But in the same mentality around differences, the first people to be accused were the homies. (All of a sudden they come wanting to change, to do things differently. They were only people who lived in the United States that had the bad luck to be deported.) For them, it's always more difficult to hold on to their jobs because [the managers] always have them in their sights.

Thus, homies may be more or less accepted in this transnational workspace, but they continue to be constructed as intrinsically violent. Homies are seen as a potentially dangerous group of workers that the company needs to control or, better yet, that the company normalizes through notions of self-control (O'Neill, 2012). Homies are only accepted if they commit to becoming "civilized," that is, by being a good worker. That means not only meeting the production standards but also adhering to the liberal prescripts of flexibility, self-development, and personal accountability.

Indeed, precisely along these lines, some homies take agency over their professional character. They incorporate and display the work ethic of the United States and the ideology of a "self-made man." Accordingly, Donald Montana claims that the deportees are more inclined to put working and earning an income over considering family and obtaining a university degree. For this reason the homies express an attitude of pride in their work, in front of the university students whose first job experience is the call centre (one reason that Vico refers to them earlier as "stupid kids"). This pride is also expressed through a youthful cultural identity in maintaining the cholo style, despite the stigmatization that it engenders. Thus, whether they realize it or not, homies at times enact the very principles of being a good worker that the firms demand of them.

However, the call centre industry takes advantage of the denigrated traits – the supposedly intrinsic violence – by encouraging deportees to use their supposed aggression for both corporate sales objectives and personal success. This could be understood as the construction of the homie as some sort of permitted "telephone gangster" who uses aggression and intensity to be competitive and reach sales objectives.

Conclusion

This article analyses the arguments for labour legitimacy on two levels. First, there is a dispute, real and symbolic, to define who is accepted in the formal labour market. This is exemplified by the almost police-like hiring practices of national companies and the exclusion of large urban populations that are stigmatized as dangerous. This set of practices naturalizes class segmentation and racialization of the formal labour market of Guatemala and operates through discourses of security and fear, which are shared in many cases by management and workers. Namely, young people of working-class origins, residents on the periphery, who are perceived as dark skinned and who wear stigmatized clothing and tattoos, are some of the most vulnerable subjects of formal labour exclusion.

In this context transnational call centres offer employment to broad segments of the population that would be excluded from formal work in other formal organizations. This employment looks modern and sophisticated, with income levels similar to those of professional jobs. However, as I have argued in this chapter, the barriers of access and the processes of exclusion continue to operate, possibly in a subtler form. They appear in the racialized stratifications of the job, practised by management and workers alike. This labour market segmentation operates through a series of symbolic references applied to different groups of workers. Homies and middle-class students are each constructed as more appropriate for particular tasks in the global service chain. Even if homies are accepted in this transnational work space, they continue to be constructed as violent characters who should be controlled or, better still, self-controlled. At the same time, the firm takes advantage of their violent tendencies, by channelling their "natural" aggression to achieve corporate sales objectives. Therefore, the homie identity is accepted *only if* workers show that they are in the process of rehabilitation and are adhering to the precepts of flexibility and neoliberal ideology by making something of themselves. This could be understood as the symbolic construction of a sanctioned "telephone gangster" who uses aggression and intensity to reach sales objectives.

Overall, the call centre industry constructs the middle-class Guatemalan worker as a civilized carrier of culture, defined by a superior education level and decent behaviours. Firms label middle-class workers, especially women, as appropriate for customer service. These workers embody the Guatemalan capacities of subservience and friendliness, which strengthen the comparative advantage of the national labour force. The practice is representative of identity management, in which call centre staff administer policies to regulate difficult labour relations inside the companies, as well as the daily interactions on the telephone between workers and their North American clients (Poster & Wilson, 2008; Poster, 2011). Even though worker-customer relations were not a direct subject of this chapter, it is important to note that Guatemalan call centre workers practise identity management on their own in many ways. For instance, workers position themselves above their clients, whom they consider "irrational," by imagining clients as inferior in terms of education mainly but also of race and culture. This is especially the case when Guatemalan workers encounter African-American or Latino-American customers. Understanding these symbolic and discursive mechanisms is a subject for further exploration.

Second, we find a symbolic dispute that is more utopian and aspirational. This occurs as workers construct new malleable identities that can adjust to the requirements of the transnational global economy. Workers face pressure in their jobs to be flexible in order to become successfully integrated into the global middle class. In Guatemala this is a symbolic struggle for proximity to modernity and to determine the groups of workers who represent best the "unbeatable – simply unteachable – degree of cultural affinity with North American consumers" (O'Neill, 2012, p. 24).

In the case of call centre work we see both legitimate and illegitimate forms of accessing this cosmopolitanism. There is a *decent* route, through formal education and access to cultural goods and global consumption. There is also a *criminalized* route, through the path of undocumented migration. This path is vindicated by its protagonists, in the adoption of the ideologies, work, and consumption patterns of North Americans. In both cases one group disqualifies the other. Homies, on the one hand, are accused of having neither the cultural behaviours nor the appropriate manners for customer service in transnational services. Middle-class youth, on the other hand, are accused of lacking a modern work culture and the ideology to make something of themselves, a fundamental requirement for work relations in the current phase of the globalization in Guatemala.

Ultimately, call centres are real and imaginary urban spaces, acting as sites of locally specific globalization strategies. Workers do not compete to be *less* Guatemalan but to be *new* Guatemalans – flexible, cosmopolitan, and consumerist. In this process they create and nourish identities that are framed by the specific moment in the trajectory of Guatemalan economic liberalization. This redefines the categories of work and gives birth to a new class of Guatemalan workers.

NOTES

1 In my previous research I collected a series of public interviews in which chief executive officers, managers, and government officials tried to refute the negative association of call centres with the *maquila* industry. See Meoño Artiga (2011), pp. 56, 90–3.
2 For example, see a newspaper column "Foro Sabatino" reporting on a parent's concern about the prevalent drug abuse in call centres. *Siglo XXI*. (2011, November 5) Los Call Centers y el uso de drogas. Guatemala City.
3 http://www.wconnection.net/Why-World-Connection.html
4 It is important to note that in Mexico the presence of deportees in call centres is largely portrayed in a positive light. Representatives of both media and academia refer to deportees as "dreamers" (Castillo 2012) and consider call centre work as an opportunity for them to begin a "new life" (*La Jornada*, 2014, 23 August). The call centre is a place in which these people can capitalize on their "binational cultural capital" (Da Cruz, 2012, p. 293). "Call centers also offer a kind of community, or cushion" in the return process (Anderson, 2013, p. 87).
5 The names of the companies in which I conducted fieldwork, and my informants worked, have been changed for the purposes of this article. Where information was obtained from public or published sources, I cite the real names of the companies.
6 According to reports by human rights organizations, more than a hundred shopkeepers are killed annually in Guatemala City. More than a thousand bus drivers, taxi drivers, truckers, and other transportation workers were killed in Guatemala between 2006 and 2012, according to the country's National Transport Commission (CNT).
7 For more references see Savenije (2006), p. 218.
8 There are currently 1.6 million Guatemalans in the United States. In 2009 they sent home remittances valuing $4.3 million. This represents 12 per cent of the national gross domestic product, which directly or indirectly benefits about one-third of the population. The remittances also represent the second-largest source of foreign exchange to the country, surpassing exports of traditional products and tourism. (Barahona, 2010).
9 SEGEPLAN, "Política Nacional de Empleo 2012–2021."
10 The names of all interviewees have been changed to protect their privacy.
11 The presence of indigenous workers in call centres is not discussed in detail in this chapter. Previous works have encountered a very marginal presence (Palencia Frener et al., 2011). Recent press releases indicate some active efforts to train and incorporate one hundred indigenous youth to

Telus. According to Telus's chief executive officer, the plan works under the idea that "an indigenous person already knows two or three languages or *dialects* and therefore has a greater ability to learn English. The program has succeeded and has incorporated to Telus over a hundred young residents of San Juan Sacatepequez" (a mostly indigenous town near Guatemala City). For more information see "Así me convertí en Guillermo Montano," retrieved from http://es/edicion12/actualidad/462/Así-me-convertí-en-Guillermo-Montano.htm; and Coronado, "El éxito es en equipo: Entrevista a Guillermo Montano" (2014).

12 Interview with Paul Briere, for the 8:45 show of Canal Antigua (2013).
13 Ibid.
14 Ibid.

REFERENCES

Álvarez Aragón, V. (2010). La educación en Guatemala: Tienda de mercancía variada. In I. Sverdlick & C. Croso. (Comps.) *El derecho a la educación vulnerado: La privatización de la educación en Centroamérica* (pp. 11-32). Buenos Aires: Fundación Laboratorio de Políticas Públicas.

Anderson, J. (2013). From U.S. Immigration Detention Center to Transnational call center. *Voices of México, 95*(03–13). Retrieved from https://www.academia.edu/3419851/_From_U.S._Immigration_Detention_Center_to_Transnational_Call_Center_.

Arana, A. (2005). How the street gangs took Central America. *Foreign Affairs, 84* (3), 98–110. http://dx.doi.org/10.2307/20034353

Barahona, D. (2010). *Migrantes Guatemaltecos en Estados Unidos y desarrollo económico local en Guatemala.* Weitz Center for Development Studies, CERUR. http://www.cerur.org/uploads/1/7/0/8/1708801/dorys_barahona_guatemala_articulo.pdf.

Caldeira, T. (1996). Fortified enclaves: The new urban segregation. *Public Culture: Bulletin of the Project for Transnational Cultural Studies., 8*(2), 303–28. http://dx.doi.org/10.1215/08992363-8-2-303

Caldeira, T. (2007). *Ciudad de muros.* Barcelona: Gedisa.

Canal Antigua. (2013, 25 November). Interview with Paul Briere, for the *8:45* show of Canal Antigua. Retrieved from http://www.youtube.com/watch?v=3P9RvGf7O38&feature=youtube_gdata_player

Castillo, M. (2012, 1 December).Deportados al call center. *Nexos,* Retrieved from http://www.nexos.com.mx/?P=leerarticulo&Article=2103065.

Coronado, E. (2014, July 20). El éxito es en equipo: Entrevista a Guillermo Montano. Guatemala City: Prensa Libre.

Da Cruz, M. (2012). *Usos de la cultura transnacional en la economía globalizada: Los estudiantes y los migrantes teleoperadores en los centros de llamadas bilingües de Monterrey (México) y San Salvador (El Salvador); En ser migrante latinoamericano, ser vulnerable, trabajar precariamente*. Barcelona: Anthropos.

Donaldson, W. (2012). Gangbangers and politicians: The effects of Mano Dura on Salvadoran politics. Stone Center for Latin American Studies, Tulane University, Louisiana. http://stonecenter.tulane.edu/uploads/Donaldson,_UploadVersion-1368207121.pdf.

Dudley, S. (2011, 15 March). InSide: The most dangerous job in the world. *InSight Crime: Investigation and Analysis of Organized Crime*. Retrieved on 4 April 2014 from http://www.insightcrime.org/investigations/inside-the-most-dangerous-job-in-the-world

Escalón, S. (2011). Magacín: En las entrañas de un call center. *Magacin 21*. Guatemala City, Guatemala.

Everest Global, Inc. (2014). *Central America and the Caribbean answer the call for English-language contact center services*. Dallas, TX: Everest Global, Inc. Retrieved from http://bit.ly/1SuyHWf.

Fleming, Peter. (2009). *Authenticity and the cultural politics of work: New forms of informal control*. USA: Oxford UniversityPress.

Fleming, P.Y., & Sturdy, A. (2011). "Being Yourself" in the electronic sweatshop: New forms of normative control. *Human Relations, 64*(2), 177–200. http://dx.doi.org/10.1177/0018726710375481

La Hora. (2013, December 28). Cifra récord de deportaciones asciende a 50 mil 221 en 2013. Guatemala City, Guatemala.

Hurtado, P. (2010, 25 July). El llamado call center. *El Periodico Guatemala*. Guatemala City, Guatemala.

La Jornada. (2014, 23 August). Mexicanos deportados encuentran nueva vida en call centers. http://www.jornada.unam.mx/ultimas/2014/08/22/mexicanos-deportados-encuentran-nueva-vida-en-call-centers-5828.html.

Levenson, D. (2011). Living Guatemala City, 1930s–2000s. In K.L. O'Neill & K. Thomas (Eds.), *Securing the city : Neoliberalism, space, and insecurity in postwar Guatemala* (pp. 25–48). Durham, NC: Duke University Press. http://dx.doi.org/10.1215/9780822393924-002

Levenson, D. (2013). *Adiós niño: The gangs of Guatemala City and the politics of death*. Durham, NC: Duke University Press. http://dx.doi.org/10.1215/9780822395621

Martel, R. (2006). Las maras salvadoreñas: Nuevas formas de espanto y de control social. *Estudios Centroamericanos, 980* (696), 957–79.

Maul, H., & Bolaños Fletes, L. (2013). *El impacto del comercio de servicios en el empleo: El sector call centre y servicio al cliente en Guatemala*. Guatemala: Organización Internacional del Trabajo. Retrieved from http://www.ilo.org/wcmsp5/ groups/public/---ed_emp/documents/publication/wcms_218869.pdf

Meoño Artiga, L.P. (2011). *Los trabajadores de la industria de call centers de la ciudad de Guatemala: Diferenciación social y representaciones sobre el trabajo*. Unpublished master's thesis, UAM Iztapalapa, Mexico. Retrieved from http://bit.ly/1Vk2xnr.

Mirchandani, K. (2012). *Phone clones, authenticity work in the transnational service economy*. New York: Cornell University Press; Retrieved from http:// www.cornellpress.cornell.edu/book/?gcoi=80140100220590.

O'Neill, K.L. (2012). The soul of security: Christianity, corporatism, and control in postwar Guatemala. *Social Text, 30*(2/111), 21–42.

O'Neill, K.L. & Thomas, K. (Eds.). (2011). *Securing the city: Neoliberalism, space, and insecurity in postwar Guatemala*. Durham, NC: Duke University Press.

Palencia Frener, S.G., Mendizábal Juárez, M.L., & Poroj Abaj, M. de J. (2011). *Género y etnicidad: De las maquilas de confección a los call centers en Guatemala; Un estudio crítico*. Guatemala City: Instituto Universitario de la Mujer de la Universidad de San Carlos de Guatemala.

Poster, W.R. (2007a). Saying "good morning" in the night: The reversal of work time in global ICT service work. *Research in Sociology of Work, 17*, 55–112.

Poster, W.R. (2007b). Who's on the line? Indian call center agents pose as Americans for US outsourced firms. *Industrial Relations: A Journal of Economy and Society, 46*(2), 271–304.

Poster, W.R. (2008). Filtering diversity: A global corporation struggles with race, class, and gender in employment policy. *American Behavioral Scientist, 52*(3), 307–41. http://dx.doi.org/10.1177/0002764208323509

Poster, W.R. (2011). Emotion detectors, answering machines, and e-unions: Multi-surveillances in the global interactive service industry. *American Behavioral Scientist, 55*(7), 868–901. http://dx.doi.org/10.1177/0002764211407833

Poster, W.R., & Wilson, G. (2008). Introduction: Race, class, and gender in transnational labor inequality. *American Behavioral Scientist, 52*(3), 295–306. http://dx.doi.org/10.1177/0002764208323508

Rivas, C.M. (2007). *Imaginaries of transnationalism: Media and cultures of consumption in El Salvador*. San Diego: University of California.

Rodgers, D. (2004). Disembedding the city: Crime, insecurity and spatial organization in Managua, Nicaragua. *Environment and Urbanization, 16*(2), 113–24. http://dx.doi.org/10.1177/095624780401600202

Romero Loyola, M. (2012). *Deportados y reclasificados: Trabajadores de un call center de la zona metropolitana transnacional Tijuana- San Diego*. Unpublished thesis for a bachelor's degree in Anthropology, Universidad Autónoma Metropolitana – Iztapalapa, Mexico.

Savenije, W. (2006). Las pandillas trasnacionales Mara Salvatrucha y barrio 18th Street: Una tensa combinación de exclusión social, delincuencia y respuestas represivas. *Intra-Caribbean Migration and the Conflict Nexus*, 205.

Wacquant, L. (2009). La estigmatización territorial en la edad de la marginalidad avanzada. *Renglones*, 60. Retrieved from http://www.redalyc.org/pdf/938/93843301.pdf

Wacquant, L.J.D. (2008). *Los condenados de la ciudad: Gueto, periferias y estado.* Buenos Aires: Siglo veintiuno editores.

PART THREE

Caught in the Middle – Labours of Borders and Crossings

7 *Migrations a l'envers*: Global Service Work and Discursive Crossings

SANAE ELMOUDDEN

As a term in popular discourse, *globalization* marks a celebrated openness of global spaces – both physical and cybernetic. The complexity of global flows demands more than the contemporary dominant binary of territorial crossing and virtual crossing. Considering offshore call centres as the epicentre of information computer technology, this chapter examines the ways in which employees in Moroccan call centres negotiate spaces through their transnational service work. Morocco's experience with globalization runs deep. Its strategic geographical location (between Europe and sub-Saharan Africa) has drawn Arab settlers and French and Spanish colonizers, who left traces of their cultures (Amazigh-Arabic), religion (Islam), and languages (French and Spanish) behind. With the advent of the Internet, Moroccan youth have embraced American influences and language. These historical preconditions enable Moroccan youth to be positioned to fill new offshore global jobs that require some digital knowledge and multi-language training. Locales such as Morocco are especially important for offshore call centre employment because offshoring organizations benefit from cheap labour and fluency of colonized languages (Hegde, 2005). Drawing on ethnographic study and informed by interdisciplinary research on space, I argue in this chapter that call centre work involves both *physical* and *discursive* crossing without territorial movement. Taking "discursive crossings" as a transgression of boundaries, the chapter illustrates the ways in which call centre workers negotiate diasporic identities in relation to both geographic and digital migration.

Introduction

In the last two decades, literature on global call centres has been abundant (Basi, 2009; Nadeem, 2011; Taylor & Bain, 2005). Building on the work of globalization and communication scholars who advocate for remapping the boundaries between the local and the global space (Massey, 2007), this chapter draws attention to the intersection of these boundaries. Drawing upon empirical data collected during 2006–7 in a call centre in Morocco, AtlasShore, this chapter explores the way in which space can serve as a conceptual tool to expand our understanding of the concept of *crossing* at the intersection of the local and the global across time.

This chapter builds on recent contributions concerning national spatial contexts in transnational service work and discusses in particular the daily crossings negotiated daily by employees in AtlasShore and which connect them to complex configurations of space, place, time, and contexts to fix the meaning of their own identities. Given the emergent and fluid quality of space provided by telecommunication technology, individuals discursively act to "fix" meanings. As in the musical concept of idée fixe, fixing is the process of selecting specific context elements (including historical and global elements) and adjusting them to fit into recurrent and identifiable themes. Thus, fixing does not create a set of enduring meaning; rather, it may shift and change in relationship to global space and the local place of the nation.

This chapter therefore aims to contribute to the call centre scholarship that engages national spatiality and place as an integral part of organizing in transnational service work rather than as an epiphenomenon (Pal & Buzzanell, 2008). In so doing, this study adds to the theorizing of the "here and now" of nationhood. Fox and Miller-Idriss (2008) argue that nationhood is made meaningful by ordinary people in the context of the everyday. I focus on the ways in which the everyday discursive construction of the local is enacted and invoked in the context of information computer technology to bridge the global space to the nation.

By using Doreen Massey's (2007) relational constitution of place and space across time, I pay attention to the intersection of *place* (the physical location) and *space* (the discursive construction of place) to strengthen the connection between scholarships of spatiality (a key concept in global service work) and the transnational call centre studies of everyday nationalism. Global geographers' studies declare that space is not empty but rather an active part of human sense-making (Massey, [1991] 1994, 2007). This tradition stresses a socio-spatial dialectic in which people are continually shaping spaces, and spaces, in turn, influence their inhabitants. That is, space is simultaneously an end in itself and a productive or reproductive process. Similarly, theorists have expanded *place* from representing a contained and fixed entity to having the notion of a boundless enclosure. One influential formulation is Massey's ([1991] 1994; 2007) concept of the "global sense of place." Her central argument is that place is unsettled and unfixed, defined in relation to "places beyond." Describing a downtown London mall, Kilburn High Road, Massey argues that the Irish Republican Army's symbols adorning the walls of the mall, alongside the stores selling British, Muslim, and Indian wares (such as saris), illustrate this place as unbounded. In the same way that IRA symbols and saris recall a place beyond, non-local symbols and European minutes in transnational call centres signify "spaces beyond."

In centres located outside of Western countries, yet serving Western clients, agents are expected to speak with and acculturate appropriate Western accents and attitudes when they are inside and outside the workplace (Perez, 2005); therefore, issues of space and place are integral to documenting the story of such work. As Renee Paulet (2008) writes, "surprisingly [this literature] has been conducted in a place-less manner" (p. 50). In her investigation of the lack of place in call centre literature Paulet argues that the location of a call centre plays an important role in its operations, policies, and practices. My study is a move towards addressing this void. The place of the study is an offshore call centre located in an Atlasian city of Morocco. In call centre work, employees (agents) sell products or provide customer service across geographical borders. The norms of daily work oblige employees to transcend the limited contours of their physical places to encounter a diversity of spatial, temporal, and contextual configurations.

Given Morocco's geographical location and colonial past, workers are a heterogeneous group composed of native Moroccans, Moroccan European citizens who have returned from European countries such as Belgium and France, European citizens who have fled high unemployment in such countries as Spain, and African sub-Saharan migrants whose original plan might be to consider Morocco as an entry point to Europe. These multicultural workers do not "cross" national borders in the same ways. For some, notions of here and there are merged when they engage in transnational service work. Others report that they feel disconnected from the local place in which they live, or in fact feel as though they live elsewhere when they engage in computer mediated work.

I will introduce the call centre AtlasShore and detail the methods used for this ethnographic study. I argue that call centre workers occupy diverse class and citizenship positions that affect the ways in which they bridge local and global interactions. Some workers have European Moroccan citizenship, and others have more limited transnational territorial mobility because they have been denied visas to travel to Europe. Some feel trapped in offshore jobs because they perceive few employment alternatives, and others idealize their jobs as opportunities to further their education or to support family members. I argue that, depending on their social locations, workers engage in different and shifting discursive crossings during their daily work. Put differently, crossing is experienced by some employees as a confinement from and into the same place, and by others as a diasporic community in which a reaffirmation of their Moroccan identity or a potential for adventure and opportunity is possible. Specifically, I offer *l'hrig* as an example of the way in which a historically shifting constellation of concepts (Adorno, 1977) may be leveraged to fix meanings of crossing discursively in transnational call centres. The concept of *l'hrig* was used in the 1970s to connote running a red traffic light (while driving) to defy systematic rules. In the 1980s the meaning of *l'hrig*, in addition to running the red light, symbolized the danger (and opportunity) of illegal immigrants' physical crossing of the Mediterranean to Europe. In the call centre, I found that the term is used to signify similar dangers and opportunities of crossing while situated in place. In this I propose that transnational studies of call centre service work consider discursive crossing not simply as affirmative (for example, generating economic prosperity from the global) or subversive (for example, generating resistance from the local), but rather as a transgression along a continuum from the local to the global. Therefore, taking the concept of crossing from communication studies that emphasize spatial contexts, I propose to understand "discursive crossings" as a transgression of boundaries within discourses and as the idea of exceeding the discursive itself by not only destabilizing the borders but also producing those same borders. The overlap of the digital and non-digital contexts in transnational service work destabilizes the conventional ideas of borders. Workers situated in the call centre of AtlasShore in the place of Morocco are able to cross (going and returning) discursively to European countries to render service to European customers, such as providing technical support.

Transnational Service Work and Workers' Identities

Call centre work is a booming business. Today there are more than 160,000 call centres worldwide, with an 11.5 per cent annual growth rate projected through 2016 (Ramakrishnan, 2011). Most call centres are characterized by a concentrated placement of desks or stations in close spatial proximity, each using communication and information technologies to facilitate interactive service (for example, sales assistance, technical expertise) for business or end-user customers. A solid body of research documents some typical characteristics of call centres, including a Tayloristic deployment of technology, codified and routinized scripts, and managerial monitoring of output (Nadeem, 2011) When call centre work "moves" offshore, issues of place take on particular importance. As Poster (2007) points out, globalized customer service work is increasing, and as a result issues of "nation" will probably continue to surface within this industry. To use Skey's (2011) more cautious approach, since "nationhood" is represented and reproduced in everyday life, it is important to pay particular attention to the local place. Offshore service work shifts service production from global north to global south. One often-debated rationale of the new "placial" shift in this industry involves savings in the costs of labour, land, and taxes. For example, monthly wages for a college-educated agent in India may be $160–$300, versus more than $2,000 for an agent in the United States (Manu, 2003).

Optimistic views on offshored call centres emphasize new employment opportunities, technical infrastructure, and skill development for call centre employees (Murphy, 2011). Transnational service work in this view has opened the global market to the less prosperous global south countries such as India and Morocco. However, the scripting of the world as a global market requires the unidirectional move from more developed countries to emerging countries at the expense of global workers' identities (Perez, 2005). As Poster (2007) describes in reference to Indian agents, the process of acting American has a few components: customizing diction and voice rhythm to mimic customers, using prepared scripts to make the trickery sincere, and adopting pseudonyms to announce American identities to global north customers. In other words, "becoming a phone clone involves emulating, through voice, an ideal transnational call centre worker who is both close to and distant from customers in the west" (Mirchandani, 2012, p. 3).

Beyond these views, transnational call centres expose the complications of understanding global work and identity, particularly as they intersect with the lived and embodied experiences of workers in space and place. Such work exemplifies what communication scholars Tracy and Tretheway (2005) identify as a dichotomy of the real self versus the fake self. Agents are valued for their ability to pretend to be not only what they are not but also *where* they are not. As Sonntag (2005) argues, global English, the language most used in transnational call centres in these instances, forms "linguistic capital" acts as a mechanism for cultural imperialism and hegemony. Despite these expectations, I agree with Tracy and Tretheway's argument that such a dichotomy masks the agentic work that individuals do in constructing positions and identities while they discursively cross between local places and global spaces through their voices: "To say that an individual is discursively constituted does not mean, however, that s/he is void of agency ... Identities emerge from hegemonic processes in the spaces between domination and resistance" (Tracy & Tretheway, 2005, p. 171).

This view reminds us that discourses are situated in place and space. For example, it would be a mistake to conclude that call centres are placeless, despite the efforts by agents to suggest otherwise to customers. Paulet (2008) examines the impact of place on three different call centres that serve the same company in Australia. Drawing on Peck's (1996) formulation of space as used by capital in search of profitable sites, and of place as used for labour constructions, Paulet concludes that factors unique to each place – including community, job market, and history – impact and influence call centre operations differently. Scholars dealing with nations instead of territories illustrate the same sentiment through ethnocentrism. For example, Mansfield and Mutz (2013) find that customers tend to have hostile reactions to agents calling them from racial groups whom they consider to be less admirable than their own ethnic group.

Similarly, in a study of a call centre in India, Pal and Buzzanell (2008) show that place and national history play a major role in the way that transnational service work is enacted and described by agents. For instance, as agents switch between pseudonyms and different accents in India, the authors observe that employees side with and adjust or readjust to different cultural expectations. Mitra (2008), for example, in using a spatial framework to investigate a call centre in India, theorizes that the new frameworks of space expose new relations of race, belonging, and colonialism and unsettle our conceptions of diasporic identities (see also Shome, 2006). Elsewhere, drawing on Roza Tsagarousianou's (2004) formulation of diasporic identities as those reproduced and negotiated in the connectivity of the cultural reconstruction of information technologies, I stress that attaining diasporic identities requires communication connectivity and willingness to belong to a diasporic space (Elmoudden, 2013). In other words, diasporic identities go beyond the limiting contours of displacement and ethnic identity found in traditional formulations of displacement requirements for diaspora (Bhabha, 1994). Studies such as these – where space and place matter – suggest new ways of understanding and interacting with spaces and places that workers inhabit in global work. Accordingly, these studies urge us to reconsider transnational theories in global service work not only regarding diasporic identities but also regarding issues of crossing.

These issues need to be examined continuously to reveal the intersections of global processes and local politics, for it is apparent that entering the global market signifies conforming to Western minutes and Western dictions. Voice and call centre work involves interactive service over the telephone, across spaces, but without territorial movement. Several dynamics, then, make discursive crossing salient within these settings. Historically, *crossing* is a term discussed in transnational studies that deal with the U.S.-Mexican border (Anzaldúa, 1999). One of the most celebrated features of crossing is the supposed enabling of free circulation and intermingling, in a world that is continuously reshaped by different meanings of globalization. This view stresses a celebratory tone of borderline fluidity (Kearney, 1998). However, in the context of collapsing borders, such as the case in call centre work, crossing should not be seen solely in relation to the celebratory tone of borderline fluidity but rather with the multiplicity of vectors associated with border crossing.

A major intervention by communication scholars demonstrates the importance of spatial contexts in globalization flows. Raka Shome (2003) emphasizes the spatial contexts at the border, claiming that an acontextual spatial focus "risks rendering invisible the situated practices of space" (p. 43). Revisiting U.S.-Mexican-border studies, Shome's central argument is that the notions of "free" and "open" have very different meanings for jet-setting immigrants than they do for transitory immigrants in the same spatial contexts. She insists, for example, that the label of *illegal* for immigrants at the border renders the bodies and the individuals at this territorial place "out of place" – belonging neither to a diasporic space nor to a transnational one.

In transnational call centres, as individuals move through the time and space of everyday life at their work, employees' spatial contextual resources of subjectivity also constantly change. Spatial contexts are important in telling the story of call centre promises and perils.

Attention to spatial contexts involves looking at places in which individuals live, and at spaces in which individuals practice their lives (Halford & Leonard, 2006). Rethinking global service work from a spatial context, I argue that crossings can be transgressive in three different ways, which are thematized here as first transgression, second transgression, and third transgression. In the first transgression discursive crossing is both stationary and mobile, and the here and there is merged through time and space. In the second transgression discursive crossing is a diasporic space in which one becomes an alien in one's own home because of the imbrication (or overlap) of the digital and non-digital communication of everyday call centre work. In the third transgression discursive crossing is inverse migration in which everyday nationalism is questionable. While not all workers undergo these transgressions chronologically, they experience them in many different instances.

The Context of Offshore Work

In the same way that countries such as India have housed U.S.-oriented transnational call centres, Morocco has become one of the most reputable transnational call centre magnets in Africa for Western countries such as France and Spain. Based on a statistical call centre study conducted by Agence Nationale de Réglementation des Télécommunications (2004) and presented to the Royaume du Maroc Premier Ministre, the country has been a leader in offshore francophone activities since 2004 owing to its proximity to Europe. According to an article written by Ali Elaalaoui (2012) in *Morocco World News*, many global companies choose Morocco for their offshore activities because they can benefit from the cheap and multilingual abilities of the Moroccan workforce and the time zone that is close to that of Europe. In April 2014 it was estimated that there were over seventy thousand call centre jobs in Morocco (Verheecke, 2014). This is a significant increase in call centre positions, considering that only ten years earlier, in 2004, fifty call centres debuted the transnational call centre imagination in Morocco, with 6,400 positions at the end of the year (Agence Nationale de Réglementation des Télécommunications, 2004). Dell Casablanca, Morocco (an American call centre for Dell), and WebHelp (a French call centre) are among the gigantic call centres that have opened their doors to Moroccan agents in big cities such as Casablanca and Rabat. Alongside these multinational call centres, a national call centre by the name of

AtlasShore opened its doors in 2004 to agents in the vicinity of the Atlas Mountains and was one of the first call centres in the region.

Research Site: AtlasShore

AtlasShore is a Moroccan-owned call centre located in the middle of Morocco that caters to francophone and anglophone markets. It specializes in selling mobile-phone memberships and delivering customer service such as computer help, assistance with cellphone questions, and giving directions. For instance, employees assist European customers by giving specific and detailed directions to restaurants or airports. Morocco has a long history of global interactions because of its proximity to Europe and sub-Saharan Africa. In addition, it has historically been a passageway for colonizers, from the Roman expansions to the French and Spanish colonizers of the last century. One outcome of French colonization in Morocco is the multilingual skills of the young and educated population. In recent years Moroccan youth have embraced British and American influences flowing from digital television and the Internet. These conditions prepare Moroccan youth well for the new offshore global jobs that require some knowledge of computers and the ability to speak multiple languages.

AtlasShore operates on two floors of a building originally constructed for apartments. The first floor is used for training, and the second floor is packed with open-space cubicles, seating sixty agents. It has grown to more than three hundred employees since the fieldwork was conducted. Among the sixty people working at the centre at the time of this study there were Russians, Belgians, French, Germans, and sub-Saharans (Senegalese and Nigerians). Most customers were located in France, Belgium, Italy, and Germany. At times customers included Americans and Canadians, and therefore employees needed to juggle time zones that differed from that of Morocco by two to six hours. At the time of this study about 58 per cent of all employees were female. Agents ranged in age from seventeen to fifty-five years old. The wages ranged from $150 to $300 per month. This was a significant amount given that an engineer's salary can start at $300 per month. For the most part the work at AtlasShore is traditional for call centre work. Employees follow the identical codified script for up to 150 calls per day in order to sell, serve, or provide technical support to their customers, although only about one-third of the calls actually extend past an initial exchange (Nadeem, 2011). However, the conditions of work are decent in that call centre workers, unlike employees in other service centres, have the ability to be creative with their dialogue and need not adhere absolutely to a codified script.

Methods

I collected data for nine months by observing the work at AtlasShore, interacting, and interviewing employees and their families. I interviewed fifty employees in total. I also interviewed the families of ten Moroccan call centre employees during the period, among them husbands, mothers, fathers, and siblings. I gathered approximately seven hundred hours of participant observation fieldwork. These observations included attending training briefings, reunions, and conferences and shadowing agents during their work and their encounters with family members. Following Lindlof and Taylor (2010), participant observation was conducted as reflexive learning as opposed to using one single observation method.

Interviews were conducted in a conversational format, following the guidelines of scholars who advocate for integrating respondents into the production of knowledge and understanding (Kvale & Brinkmann, 2008). Interview questions often arose out of the conversations with the respondents. Interviews were also contextual in that they explored observations in depth. I said, "Tell me what you were thinking to yourself when you were providing directions to your customer in Brussels." The open-ended, semi-structured format helped to guide participants to discuss narratives of their experiences and gave them room to bring up any unanticipated elements of their daily activities (Lindlof & Taylor, 2010). These interviews ranged from thirty minutes with agents (time was limited to lunch breaks) or forty-five to sixty minutes with family members. Interviews were recorded if permitted and then transcribed.

I conducted data analysis using the three stages of grounded theory (Creswell, 2007): open coding (identifying categories), axial coding (interconnecting categories), and selective coding (choosing a core category to which most categories interconnect). During the open coding stage I categorized codes based on data comparison collected from transcripts and field notes. In the axial coding stage I diagrammed data in nodes where relationships between categories were identified as causal (associations via cues such as *when, because, is*), contextual (when and how many times), and interventional (history, culture, society). Categories relate to each other in causal and contextual manners, such as, "We are Arab, we have a history of sales; that is why the call centre is in

Morocco." Finally, I used the selective coding stage to provide a story in which the main category pulled coding from the other categories (open and axial). It was in this stage that the Moroccan word *l'hrig*, which means "crossing," surfaced as a term used by people inside and outside the call centre, and hence it became a central category in the analysis of the call centre which follows.

Interpretation of Results

The colloquial Moroccan Arabic word *l'hrig* comes from the classical Arabic *hariq* and the verb *haraqa*. Translated into English, these words mean "to burn" (Wortabet, 1995). Prior to the 1990s, *l'hrig* was associated in Moroccan Arabic (known as Darija) with being burned by fire (or burning a passport in fear when crossing illegally) or "burning a red traffic light" (that is, driving across an intersection when a traffic light is red, hence defying systematic rules). Since the early 1990s the newer meaning of the word *l'hrig*, in addition to "burning," refers to "crossing" the visible Moroccan borders by invisibly traversing the Mediterranean Sea to reach the West. *Harraga* (singular, *harrag*) is the name given to the people who accomplish the daring act of *l'hrig* by illegally immigrating to Europe. It is about leaving one's nation to search for the El Dorado (a metaphoric land of gold in fiction and myth, applied in this case to describe where workers can find the great job, the money, the cars, and ruminations) in the host European nation. At times this search is unattainable because of the risk of death on the small boats used to cross the sea.

Doing *l'hrig* means traversing the fourteen kilometres of the Strait of Gibraltar, squeezed among many bodies in small boats, where the success of the journey is uncertain and unpredictable. Prior to embarking on crowded boats, travellers save or borrow thousands of dollars to pay human traffickers. Between 2006 and 2008 fifty thousand clandestine immigrants landed on the shores of Italy, 60 per cent of whom were people from North Africa (Sinibali, 2008). According to the Italian government, in 2014 alone twenty-two thousand clandestine migrants reached Italy by boat, which is ten times more than during 2013 (Kassar & Dourgnon, 2014). *L'hrig* in this context is not only a very daring act because it defies the Spanish, French, or Italian authorities; it is also a very perilous act, which denotes either a real courage or a profound despair, or both. As expected, the daring act of risking one's life to flee the country through illegal emigration by land has become prevalent in Moroccan fiction (Smolin, 2011).

To date, more than thirteen Arabic and eighteen francophone novels and short stories describe the different themes associated with this practice (Jelloun, 2006). The authors depict the cruel reality of *l'hrig*. Some recount the contact with the Moroccan migration mafia and the economic costs of the crossing. Others describe the storytelling and images of Europe as an "El Dorado" (the place of good jobs) that compel characters to face the harsh conditions of crossing. Perhaps this phenomenon is most familiar to Americans from a novel by Laila Lalami, *Hope and Other Dangerous Pursuits* (2009). In essence, I want to bring the understanding of such a word to academic writing, hoping to bring to the forefront the potential implications of transnational service work in a non-Western environment.

Today, in addition to the idea of physical crossing, a new meaning of *l'hrig* is being coined by workers in transnational call centres in Morocco, referring to a discursive crossing. In telling the story of AtlasShore, I will go through the daily discursive crossings of employees in an offshore call centre in a traditional Moroccan city. To tell their story, I will present the call centre through a vignette to invite the reader to experience the participants' daily discursive crossings and diasporic identities.

One participant is Marianne, whose real name is Soumia. She took this alias when her boss asked her to choose a European name to accommodate European customers. She reveals the promise and perils of discursive crossing in a workplace that mixes the global and the local. "My new name, Marianne, was destined to me; it came to me automatically when [call centre managers] asked us to choose a French name in the centre. As you may know, Marianne is a woman who represents the symbol of liberty in France, and at the moment of my introduction to this job this name came to me," says Marianne when she is asked about her choice of pseudonym. Soumia was born in France of Moroccan immigrants and returned to Morocco after marrying a Moroccan man. She shares this trait with a few call centre workers (about a quarter of the employees at the time of this field study): the enjoyment of both French and Moroccan citizenship. After having three children, Marianne was looking to be "an independent woman," but as she was proficient in French alone (she cannot speak Arabic), she "had to wait until the call centre came to Morocco." *Marianne*, as a chosen pseudonym, is interesting. Marianne is the image of equality in France and is powerful when used in this context. She was a very popular figure in the seventeenth century among the women of the kingdom of France, in the global north. Her name later became the symbol of liberty, equality, and fraternity.

In the twenty-first century, although this symbol still represents France, it is depicted by a model of a woman who is usually young, strong, and calm, which are all characteristics needed to face work at the offshore call centre. To answer up to 120 calls a days (Thompson, 2003), repeating the same codified script imposed by their employers, agents need to learn the virtue of patience. Soumia started as an agent, then became a trainer, worked her way up to being a supervisor, and is now finally a manager. Her professional transformation went from acting like the global north to befitting the global north in a short amount of time. Early in her career she was servicing the West as an agent by conducting customer satisfaction surveys or selling cellphone memberships, and now, as a manager, she is hiring agents who are Moroccan citizens (seeking opportunities and adventures in the call centre), sub-Saharan African migrants (hoping to transition to Europe through Morocco), or Westerners (fleeing European economic crises). It is important to note that many of the stories in this chapter are not Marianne's alone but rather an amalgamation of several employees' stories, to illustrate the three transgressions of discursive crossing discussed earlier. Therefore, integrating the image of Marianne throughout the analysis seems fitting.

Analysis: Constellation of Concepts and Transgressions

As might be expected, the participants interviewed for this study support the kinds of values typically associated with a transnational call centre (Brophy, 2011). Employees discussed the advantages of call centre work, such as new employment opportunities, technical infrastructure, good pay compared to that of other jobs in the country, and skill development (Murphy, 2011). However, similar to Pal and Buzzanell (2008), I find that intersections of space, place, and time play a major role in the way that work is enacted at AtlasShore. Employees expose the complications of understanding "crossing," particularly as they intersect with the lived and embodied experiences of space and place. Participants are valued for their ability not only to pretend to be what they are not but also to pretend to be *where* and *when* they are not. The findings suggest a shifting meaning of *l'hrig* as a constellation of concepts associated with the call centre.

Coined by Theodore Adorno (1977; see also Benzer, 2011), a "constellation" is an assemblage of equally ranked concepts that are configured in different ways, based on logical inference. I extend this concept to include a reformulation of these configurations based on time and space and through communication technologies. This constellation of concepts is a result of understanding place as a continuum of layers of the physical location (for example, AtlasShore, Atlas, Morocco, and the globe), while global space is taken as a discursive construction of place. In this sense, the phrase *constellation of concepts* signifies a cluster of meanings that shift and change across place and time. These shifts coalesce with the historical meaning of *l'hrig* (crossing), not only *physically* (that is, from the south to the north) but also *discursively* (in many directions). Discursive crossing in this sense defies the subversion or affirmation associated with discourses of globalization and shows employees' daily transgressions.

In what follows I discuss the three different transgressions. It should be noted that discursive and physical negotiations are complex and in many ways incomplete. Nevertheless, such formulations are important if we assume that, in the celebratory tone of transnational fluidity, borders are not only destabilized but also reproduced socially at the physical locations (Shome, 2006).

First Transgression (Discursive Crossing as Both Stationary and Mobile)

The daily work of AtlasShore workers depicts agents rushing to their cubicles in the *marguerites*, daisy-like seating arrangements equipped with six computers and telephones, and surrounded by glass walls. Agents systematically put on their headphones and log on to their computers to begin their day. Their computers all show 10:00 a.m. (local time in France, which is the equivalent to 8:00 a.m. local time in Morocco). The humming of the agents' voices echoes across the "production plateau." Most agents are buzzing in French, a few are using Spanish, fewer are conversing in English, and two agents are speaking German. Voice-embodied smiles in all these languages are entering homes in Europe, the United States, and Canada. This initial picture of AtlasShore exemplifies the here and now of the everyday context in which nationhood is made meaningful (Fox & Miller-Idriss, 2008). Located in Morocco, agents are able to experience multiple temporal and spatial dimensions from the local place. The daily interactions expose a complex interplay of colonial histories, gender practices, and national interests (Mirchandani, 2012). Men choose call centre work because they are students or college graduates who have not find better opportunities. Although this true for women as well, some women see this type of work as emancipation from housework.

When I first met Soumia at the call centre, she was an agent. When AltasShore moved to Morocco, it allowed her to use her French and achieve some freedom from housework. During his interview Christian (a Moroccan-Belgian returnee) expressed a similar sentiment. Like Marianne, Christian has affiliations with both Belgium and Morocco because he was born and grew up in Belgium, and his immigrant parents came from Morocco. He relates, "I always wanted to go back to Belgium, but now Belgium came to me since I cross over daily." This sentiment is also shared by Sylvie in her explanation of her job as "a bridge, a sort of extension and connection to Europe." While these workers conceptualize *l'hrig* as a journey or an escape, others refer to associated meanings of the term such as "daring," "hope," "risk," or "danger." Specifically, David's comment illustrates the act of daring: "Sometimes I get someone on the phone that detects my accent and tells me he knows I am in North Africa and that he had heard of 'us,'" says David during his interview. "To which I reply, 'My nationality is French of Arab descent. Do you have a problem with my origins?' This calms them down, usually."

Samuel emphasizes "the shrewdness of Arab salesmanship" and "the kindness of the Moroccans to Europeans" as reasons for the success of call centres in Morocco. Julia, another agent reflecting on the conditions of her crossing, adds, "I talked to a customer in France who asked me where I am located. Not believing my answer, the customer says, 'It is okay. People in France know that call centres are located in Morocco and that Moroccans (and not French) are calling them. That is okay because the French cannot sell, and Moroccans are known for their Arab trade skills.'" Samuel, a thirty-five-year-old agent, adds, "You know why there is all this delocalization? It is because they know that we, Moroccans, are good to foreigners. We are used to the French from the days of colonization and we respect them. Their agents are not as patient as we are, nor are they as shrewd when it comes to selling. We are Arabs. We are Atlassy (people from Atlas). Trade has always been our strength." Michael, later, re-emphasizes: "Morocco is the centre of the world. The Atlas is at the centre of Morocco, therefore the Atlas is at the centre of the world. That is why we are very good at languages, and our culture is a syncretism of everything. Look, French and English are languages that I use to speak to my Internet friends. You see we have family everywhere, we can adapt to all situations. That is why they [Westerners] like to have call centres in Morocco." Issues of nation-hood, colonialism, and multilingualism are evident in such discursive construction of the local place. While most inhabitants claim Moroccan citizenship, they differ in terms of ethnicity. The CIA's *World Factbook* describes Morocco's ethnic-group as 99 per cent Arabic-Berber, but people refer to themselves as Arabs or Tamazight. Another insightful example is clear in Julia's interaction with her customer who indicates that the French know about the little secret of call centres in Morocco, which are there because "Moroccans are known for their Arab trade skills." In fact, drawing on the past and place is a key factor in training. "Oh yeah, you know, in training they tell us that we, the Moroccans, are tradesmen and we know how to sell better to Europeans," says James.

These agents seem to fix the meaning of *l'hrig* within the nation of Morocco into a crossing. Beyond the action of crossing a traffic light, *l'hrig* is the action of traversing the distance, fourteen kilometres of seawater, to reach "El Dorado." The promise of "landing" good jobs in Europe and bettering one's life remains in people's consciousness. Immigrants returning with beautiful cars reinforce this image. Given that about 30 per cent of Moroccan youth are unemployed (World Bank Report, 2012), *l'hrig* becomes something to which many youth aspire and practice.

The sense of crossing deepens with new associations of leaving and expectation of return for these call centre workers. For example, Jessica, one of the supervisors and a French-Moroccan returnee, logs off all the computers (which show 9:00 p.m.) and says, "It is 7:00 p.m., it is time to go home." She continues, "I like being here. Every time I leave the cen-tre, I feel as if someone is taking my passport away from me and I will have to re-enter Morocco." The examples of Jessica (conceiving of the symbolic "passport") and Christian (making sense of his work as crosso-ver to Belgium) are useful. Agents emphasize a global articulation of the physical place of Morocco. The symbolism of the passport is powerful not only in terms of crossing but also in terms of the nation. A passport is the legal document of entry to the global north. The call centre provides a bridge of legal entry to Europe. So, in a way, the sneaking into France happens without the need of the legal document.

The bridging in AtlasShore is manufactured through the digital manipulation of time and space to match what exists across the Medi-terranean. For instance, computers show 6:00 p.m. (4:00 p.m. Moroccan time). In Europe, 6:00 p.m. is the time at which people re-enter their homes, and therefore it is the time for agents to make the most sales.

But first it is the time for new energy with tea and pastries. All humming stops for ten minutes, and everyone follows the smell of mint tea coming from the kitchen. Although the time for tea in Moroccan homes is 6:00 p.m., agents drink it at 4:00 p.m. Moroccan time to match the European tea time.

Yet *l'hrig* is dangerous and risky, achieved by crushed bodies in small boats over the fourteen kilometres. Call centre agents, similar to the *harraga*, are "trapped" and imprisoned at their workplace during a daily journey that has no guarantee of a successful future. Contrary to Christian's and Sylvie's "choice" to stay and enjoy "the bridge," Mona quits "the prison" of offshoring. Mona feels trapped in the call centre with no way of advancement. The organizational ladder is very limited in the call centre. While she is satisfied with the current pay, she relates that few agents become supervisors, and even fewer workers can become managers. Faced with a similar situation, Daniel gears his attention to the stationary spatial area of the call centre where he is in a job that offers good money but no future. Many other agents stay confined within the incarceration of offshore work, a motionless aspect of offshore, benefiting from the "good pay" of the call centre but "trapped in small spaces" and restrained from "speaking Arabic," one of the native languages. The fixing of meaning by these agents is vast and important. Even though they are faced with similar physical places and contradictory contexts, the agents respond differently. The multiple discursive construction of place via the nation, the city, and the organization emphasizes the notion of movement and stability at the same time. The constellation of *l'hrig* assumes the daring, the trap, the risks, and the benefits of traversing or staying within a place (see figure 7.1).

In these examples discursive crossing is both stationary and mobile where the here and there is merged through the time and space of the digital and non-digital overlap of the call centre work. The discursive crossing of moving and staying seems to grant agents two citizenships. They are able to work legally in a global space, while accessing the continuum of the local place: from the organization, the city, and the nation to the global spaces of the construction of the nation of Morocco. In this transgression, crossing is both affirmative and subversive in that workers transgress laws and visas, overcoming institutional regulations, by clandestinely migrating across national borders from a stationary place in the call centre. This leads to my second finding concerning the discursive crossing as diasporic space.

Figure 7.1   Constellation of *l'Hrig* as Crossing the Traffic Light.

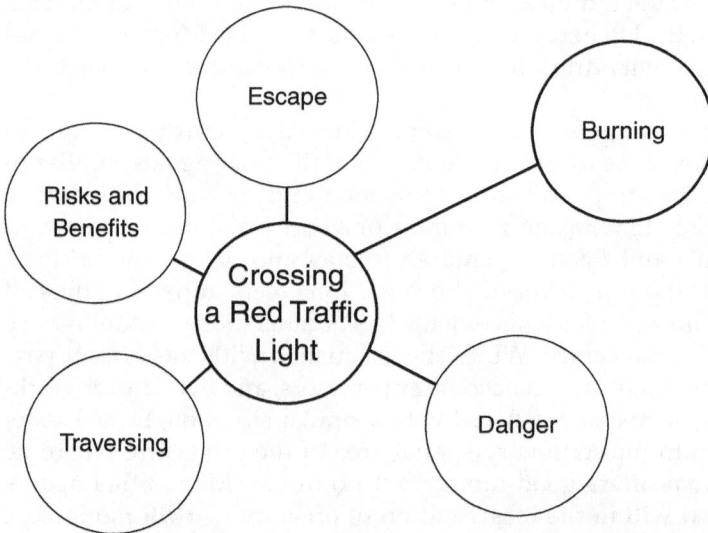

Second Transgression (Discursive Crossing as Diasporic Space)

When I was presenting a talk about Marianne at one of the local universities during my stay in Morocco, an audience member conveyed, "You
know, you can say I am one of the sons of Marianne. My mum was one of
these people whose parents emigrated to France and was born in France.
After she married a Moroccan, my parents returned to Morocco and gave
birth to me. I grew up with French in the house and Arabic in the streets.
Now I am a student at the local university, and because of my proficiency
in French I am able to support my studies through the call centre job."

Although this discursive construction is minimal, "the sons of Marianne" is an interesting representation of the contradictions of discursive
crossing. It moves the idea of crossing beyond something that is only
either affirmative or subversive and introduces the notion of diasporic
space. During their half-hour break I followed the workers to the lunch
room where I observed a conversation sparking between Marianne and
her colleagues about the blessings and the risks of their global work.
Mary Jeanne, an agent ("one of the daughters of Marianne") who is
finishing her bachelor's degree and took this job to finance a large portion of her studies, says, "It is a life boat or a life jacket." Daniel refutes,

"I came back to work here even though I feel imprisoned here, but you see, once you get used to offshore money, everything else is peanuts." "I am so tired of repeating the same things all day long in French. This is like a prison. I have to leave this place," says Mona in the cramped lunch room. "You should have some beer and forget your troubles," says a supervisor. The conversation continues:

A FEMALE AGENT, *jokingly*: As the doctor says, one glass of wine a day is good for your health.

A VEILED AGENT, *relaxing on the sofa*: Is that a Muslim doctor?

JEFFERY: Who cares if she or he is a Muslim doctor? Anyway, where do you see any Muslims here? I see only Françoises, Jacquelines, Jefferys, and Michelles.

A FEMALE AGENT, *dressed in tight jeans and a sexy top*: Well, it does not matter, we are still in Morocco.

JACQUELINE: Yeah, but here we are in a different place. There are all these people who are French.

I follow Marianne to the training centre, who says that "it is time for training the new agents." Marianne, the figurative mother who now holds a new job (instructor), says, "Let's practise our French accents some more, and in doing so we can debate the benefits of the call centre in Morocco." Eighteen-year-old Dana shares: "You see this [call centre] is changing things – it is interesting. In Europe when people do not want to do a job, they say, 'Go do the job of an Arab.' Now we [Arabs] are teaching them technology." Samuel, one of many at the centre who has been refused a visa to enter Europe, emphasizes "the shrewdness of Arab salesmanship" and "the kindness of the Moroccans to Europeans," a residual of colonization. David, who was also denied entrance to Europe, says, in thinking about his daily telephone calls at the offshore work, "you see, in a way, we are like immigrants in Europe. You know that there are many Arabs with different accents who live in Europe." In these cases and others, where legal migration on land is out of reach, legal migration via global spaces within information communication technology is attainable. Damian, who has an "all but dissertation" (meaning partial fulfilment of a doctoral degree) in French literature, disagrees: "The centre is *un chomage déguisé* [a disguised unemployment], a temporary job, a trapping prison, until something else comes along." The word *chômage* in the French language means "unemployment," but for Moroccan immigrants it means both unemployment and the social aid given to the unemployed in France. It is brought back to Morocco through immigrant (including *harraga*) discourses. Legal immigrants who have been to France bring back the notion that if one is unemployed, one may still receive money from the state in the form of entitlements (a notion that is unthinkable in Morocco). This money is called *chômage*. By relating *chômage deguisé* to the call centre, Damian is drawing on immigration and, significantly, the trap of the welfare system in Europe after territorial crossings. During job lay-off periods some immigrants end up drawing on the welfare system in Europe to make ends meet while aspiring for another job. In this sense the call centre becomes a trap, a sort of *chômage deguisé* where agents aspire for another job with less work hours and more stability. For instance, in an article about call centre unionization in Morocco, Ayoub Saoud, general secretary of the trade union representing workers at B2S Morocco, clarifies that many workers are laid off before they reach the pay-rise mark of two years after being hired (Verheecke, 2014).

Making such configurations of space, place, and time, these agents relate the call centre to *l'hrig*. Yet, this is a newer constellation of concepts surrounding *l'hrig*. If risk in other contexts has meant the "danger in traversing," here in the call centre it means the "danger of becoming an alien in Morocco." A pointed example is the conversation that took place at lunchtime regarding drinking and Muslims in Morocco, described above. Jefferey insists that in AtlasShore "there are only Jacquelines, Françoises, and Christians," and Jacqueline says, "Yeah, but here we are in a different place. There are all these people who are French." If outside the call centre there is an expectation of bodily return through *l'hrig*, inside AtlasShore the body is returned and still crosses on a daily basis. Another example that is representative of this reconceptualization of *l'hrig* is the discussion between Christian and his friend: "Shno S'hablik Mazal Fi Fransa, Rak Fi Lmaghrib daba. Hragti we rja'ti" (Do you think you are still in France? Now, you are back in Morocco. You sneakily crossed to France and came back). According to Christian's friend, *l'hrig* (or the sneaking) is associated with the physical crossing. In the original understanding of *l'hrig*, the illegal entry into territorial Europe is shameful because it represents sneaking into a country that despises illegal immigrants. But *l'hrig* here can also be interpreted as the cleverness of sneaking into Europe illegally (local defiance of passport) and legally (global acceptance of transnational workers) at the same time.

While Daoud's following passage might emphasize the new-found technological equality between Arabs and Europeans, the main take from his words is the creation of a new diasporic space in the context of *l'hrig*. "Why leave the sun of the Moroccan Atlas? This job changed my vision about the French. I used to think that the French are beyond us, but now I see that we help them even in technical support. Pourquoi je vais 'nahrag' pour longtemps quand je peux 'lahrig' chaque jour?" (Why should I cross illegitimately to Europe for a long period when I can cross legitimately every day?) The first level of interpretation of this excerpt looks at equality with the Europeans and economic prosperity. *Harraga* (those who dare to commit this act, in other words, cross to France) can return to Morocco and become economically equal to Moroccan middle class. A deeper level of interpretation, however, shows that Daoud is able to create a space for himself where diaspora is a daily occurrence and a continuous process that extends from the local place of the call centre, through Morocco, to the global. The call centre not only opens a diasporic space for agents who can avoid moving physically in search of prosperity elsewhere, but also keeps them confined in place.

The diasporic space happens outside the call centre as well via the media, when administrators insist on agents watching French television news (see for instance Shome, 2006, for similar documented instances of the use of media in other offshore call centres such as those in India). In discussions about the call centre with workers' family members, references to *haragga* were made a few times, as in the case of Victoria's husband who said, "Call centre employees are *haragga*," and, "Although it might take them a little while when they leave the centre to think they are back in Morocco, when they see a Moroccan bazaar, they know they are back." In thinking about the call centre, Victoria concludes, "Perhaps, it is both a curse and a blessing." Subsequently the constellation of *l'hrig* adds a cluster of immigration (with its own associations such as "alien" and "diaspora") to its prior collection of daring, risk, trap, opportunity, etc. Therefore, the constellation of *l'hrig* adds a new meaning of diasporic spaces for agents like Marianne and "her children" while they are located in the same place of Morocco (see figure 7.2). The discursive crossing comes as a diasporic space in which one becomes an alien in one's own home because of the imbrication (or overlap) of the digital and non-digital communication of everyday call centre work. The transgression in discursive crossing in this sense happens in the continuum of place – from organization to city, to nation, and to globe.

Figure 7.2 Constellation of *l'Hrig* as Fourteen Kilometres.

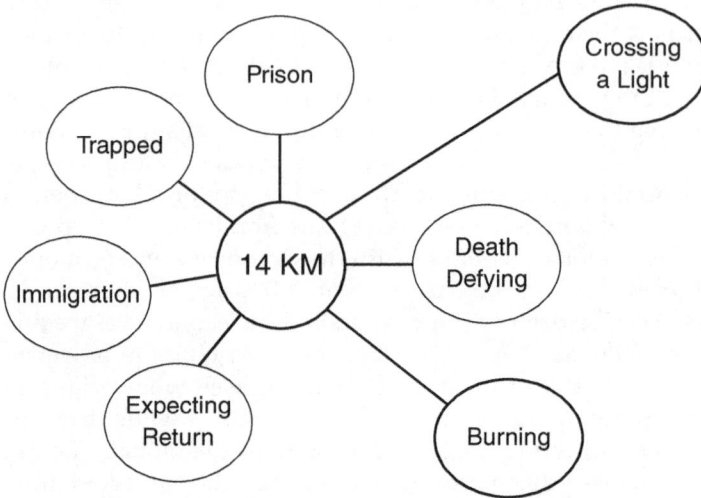

Third Transgression (Discursive Crossing as Inverse Migration)

Beyond the discursive crossing that individuals do when constructing diasporic spaces in the continuum of place, there is a new-found power at the local level. This is illustrated by Moses: "You see, I do not think I want to leave the centre. Look, we support technology in the multinational. You see, this [call centre is] changing things. It is interesting in Europe, when people do not want to do a job, [they] give it to an Arab. Now we, in Morocco, are showing them how to do things. Not just anything, we do technology. In the call centre we joke, saying, 'This is the end of life as we know it, [the] third world is more knowledgeable about technology than [is the] first world.'" This sentiment indicates that *l'hrig* in the form of discursive crossing is changing power relations to some extent between the south and the north.

The sentiment is further explored in the management voice of Marianne (who became a manager within a year of the field study), with whom I followed up after my fieldwork. She described a trend that I refer to as "inverse migration": "You know, now the French come seeking jobs in our call centre. The number of employees has increased from 60, when you were here, to more than 300 employees. In a way, they are *hragging au Maroc* [they are crossing over to Morocco]. LOL. They come asking me for jobs because of the global economic crises in the West ... You can say it is *l'immigration à l'envers* [the immigration in reverse]. Now we have the El Dorado, and why not, the sun is beautiful in Morocco." This excerpt suggests that in the interplay of the local place and the global space there is a hint of a shift in power relations. The call centre is not only about servicing the West but also about producing jobs for local and international employees, be they sub-Saharan immigrants or European transients.

Transnational centres provide work to Moroccans and also to Europeans who come searching for work in post-colonial places like Morocco. Crossing here violates the unspoken but commonly assumed law of globalization in terms of a unidirectional migration from south to north. *L'hrig* is happening in the inverse sense, from Europe to Morocco. For instance, Hustad (2013) articulates in "'The World Upside Down': The Rise of Spanish Immigration to Morocco" that the number of Spaniards has quadrupled in Morocco owing to economic instability in Spain. Both Europeans and Moroccan-born Europeans have migrated to Morocco, seeking jobs. According to the the national statistics institute INE, Spain, the share of Spaniards who left Spain seeking work elsewhere, for instance, increased by 8 per cent in 2012 from 2008. At the same time, the number of Spaniards moving to Morocco increased by 32 per cent from 2008 (Benoit & Karam, 2013). Many try to find jobs in transnational centres such as AtlasShore (Yabladi.com, 2013). Significantly, the call centre provides these new immigrants with ways to keep in touch with their homelands. It gives them a chance to cross discursively, while doing things like providing customers with precise directions to restaurants, as if they resided locally. Jean Claude, a French citizen, tells me that his wife is a Moroccan returnee, and he decided to work at the call centre. "I like being here because it is like we are in France. The facility is equipped with central air conditioning. I spend all day speaking in French with French people."

Given the many configurations of space, time, and place that agents face at AtlasShore, agents make different associations with the cluster of meaning of *l'hrig*. In a way, the constellation of *l'hrig*, beyond crossing in one direction, adds a cluster of *immigration à l'envers*. Put differently, discursive crossing is another form of transgression that violates assumptions of traditional globalization relating to the transition from the south to the north (see figure 7.3).

Figure 7.3 Constellation of *l'Hrig* as Call Centre.

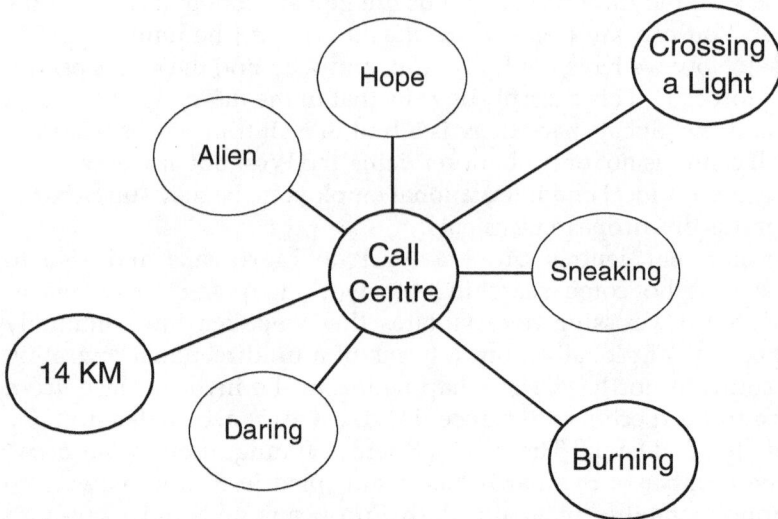

Hope

Crossing a Light

Alien

Call Centre

Sneaking

14 KM

Daring

Burning

Discussion and Concluding Remarks

In telling this story I do not deny other types of workforce exploitation associated with transnational call centres (Shome, 2006) such as acculturation of Western values, high-pressure call volumes (exceeding 150 a day, depending on the day or the shift), and emotional stress (Thompson, 2003). In fact, as recently as April 2014, thousands of call centre workers took to the streets of Morocco to protest the long hours of call centre work and demanded trade union rights (Verheecke, 2014). However, I suggest that a focus on the interplay of the global space and the local place may give us a clue to the deeper socio-politics and constructions of nation in call centre work, especially in the south. Complex configurations of space, place, time, and context are increasingly prevalent within transnational call centres, and hence constructions of nation will also increase. By "questioning the categories we normally take for granted" (Taylor & Van Every, 1999, p. 65), this study advocates for place and space as analytical tools for understanding meaning in these globalized organizations. This departs from the traditional privileging of either the local or the global and instead emphasizes a framework of place and space in transnational call centre work.

My research question deals with the fixing of meaning in physical places, across time and space, in transnational call centre service work. Drawing on fields of geography and communication, I conceptualize space as multiple, dynamic, and contextual ([1991] 1994). I borrow and elaborate on a conceptualization by Adorno (1977) and suggest that call centre agents are drawing on a "constellation of concepts" – a changing configuration of ideas, dynamic in time and space – that create meaning for their work by association with their understanding of the past and their place. These coalesce for AtlasShore workers in *l'hrig*, a constellation of concepts that includes such notions as daring, trapping, hoping, and burning.

Moreover, the constellation of *l'hrig* reveals multiple meanings infused with contradictions. Under the call centre's meaning of *l'hrig*, "crossing" is simultaneously *movement* and *stasis*. The "sneaky" crossing to Europe, represented in this chapter, in essence creates a discursive space of immigration to Europe. Contrary to Smolin's (2011) explanation of *l'hrig* as "burning the past," I bring nuance to *l'hrig* by reframing it as a *legal* discursive crossing. The crossing within this offshore call centre is a daring to leave Morocco *without burning* passports or burning the past. In fact, *harraga*, in the call centre, are those who can negotiate their identities by *both* leaving *and* staying in Morocco at the same time. They reframe their identities in the physical place of the call centre in order to cross over globally via the multiple spaces and technologies of transnational service work. By highlighting discursive crossing as a transgression of boundaries, this chapter illustrates the way in which respondents shape diasporic identities in relation to both physical migration and discursive immigration. Diasporic space, therefore, exists not only within the travel across material geography but also within communicative interactions in the networked economy. As seen in this account, the constructed "nation" is at the intersection of the local to the global spaces.

REFERENCES

Adorno, T. (1977). Commitment. In *Aesthetics and Politics* (pp. 180–91). London: New Left Books.

Agence Nationale de Réglementation des Télécommunications. (2004, October). Synthèse des résultats de l'étude realizé par CSC Peat Marwick / Capital Consulting – ANRT du développement des centres d'appels à l'émergence du BPO au Maroc. Retrieved 30 October 2015 from http://costkiller.net/tribune/Tribu-PDF/callcenter-ANRT-maroc.pdf.

Anzaldúa, G. (1999). *Borderlands / La frontera: The new mestiza*. San Francisco: Aunt Lute Press.

Basi, T.J.K. (2009). *Women, identity and India's call centre industry*. London: Routledge.

Benoit, A., & Karam, S. (2013, August 28). Spaniards fleeing jobless scourge seek jobs in Morocco. *Bloomberg Buisnessweek*. Retrieved 13 April 2016, http://www.bloomberg.com/news/articles/2013-08-27/spaniards-fleeing-jobless-scourge-seek-jobs-in-morocco-economy.

Benzer, M. (2011). *The sociology of Theodor Adorno*. New York: Cambridge University Press. http://dx.doi.org/10.1017/CBO9780511686894

Bhabha, H. (1994). *The location of culture*. London: Routledge.

Brophy, E. (2011). Language put to work: Cognitive capitalism, call center labor, and worker inquiry. *Journal of Communication Inquiry, 35*(4), 410–16. Retrieved 28 March 2014. http://dx.doi.org/10.1177/0196859911417437

Creswell, J. (2007). *Qualitative inquiry and research design: Choosing among five traditions*. London: Sage.

Elaalaoui, A. (2012, 17 January). The growing call center industry in Morocco. *Morocco World News*. Retrieved 28 October 2015, http://www.moroccoworldnews.com/2012/01/23053/the-growing-call center-industry-in-morocco/

Elmoudden, S. (2013). Moroccan Muslim women and identity negotiation in diasporic spaces. *Middle Eastern Journal of Culture and Communication, 6*(1), 218–37.

Fox J.E. & Miller-Idriss, C. (2008). The "here and now" of everyday nationhood. *Ethnicities, 8*(4), 573–6. http://dx.doi.org/10.1177/14687968080080040103.

Halford, S., & Leonard, P. (2006). Place, space and time: Contextualizing workplace subjectivities. *Organization Studies, 27*(5), 657–76. http://dx.doi.org/10.1177/0170840605059453

Hegde, R. (2005). Disciplinary spaces and globalization: A postcolonial unsettling. *Global Media and Communication, 1*(1), 59–62. http://dx.doi.org/10.1177/174276650500100114

Hustad, K. (2013, 21 March). "The world upside down": The rise of Spanish immigration to Morocco. *Christian Science Monitor*. Retrieved 15 January 2015 from http://www.csmonitor.com/World/Africa/2013/0321/The-world-upside-down-The-rise-of-Spanish-immigration-to-Morocco.

Jelloun, T.B. (2006). *Partir*. Paris: Gallimard.

Kassar, H., & Dourgnon, P. (2014). The big crossing: Illegal boat migrants in the Mediterranean. *European Journal of Public Health, 24*(suppl. 1), 11–15. http://dx.doi.org/10.1093/eurpub/cku099

Kearney, M. (1998). Transnationalism in California and Mexico at the end of empire. In T.M. Wilson & H. Donnan (Eds.), *Border identities: Nation and state at international frontiers* (pp. 117–41). Cambridge: Cambridge University Press. http://dx.doi.org/10.1017/CBO9780511607813.005

Kvale, S., & Brinkmann, S. (2008). *InterViews: Learning the craft of qualitative research interviewing* (2nd ed.). Thousand Oaks, CA: Sage.

Lalami, L. (2005). *Hope and other dangerous pursuits*. Chapel Hill, NC: Algonquin.

Lindlof, T., & Taylor, B. (2010). *Qualitative communication research methods*. Thousand Oaks, CA: Sage.

Mansfield, E., & Mutz, D. (2013, September). Us vs. them: Mass attitudes towards offshoring and outsourcing. Paper presented at annual meeting of American Political Science Association, Seattle, WA.

Manu, J. (2003, 11 June). IT sweatshops breaking Indians. *Tech Biz*. Retrieved 20 October 2014 from http://archive.wired.com/techbiz/media/news/2003/07/59477

Massey, D. ([1991] 1994). A global sense of place. *Marxism Today*, 24–9. Reprinted in Doreen Massey, *Space, place, and gender* (pp. 146–56), Cambridge: Polity Press.

Massey, D. (2007). *For Space*. London: Sage.

Mirchandani, K. (2012). *Phone clones: Authenticity work in the transnational service economy*. Ithaca, NY: Cornell University Press.

Mitra, A. (2008). Quality of life of call center employees in India: Preliminary findings. In J. Luca & E. Weippl (Eds.), *Proceedings of world conference on educational multimedia, hypermedia and telecommunications* (pp. 6290–6). Chesapeake, VA: Association for the Advancement of Computing in Education.

Murphy, J. (2011). Indian call center workers: Vanguard of a global middle class. *Work, Employment & Service, 25*(3), 417–33.

Nadeem, S. (2011). *Dead ringers*. Princeton, NY: Princeton University Press. http://dx.doi.org/10.1515/9781400836697

Pal, M., & Buzzanell, P. (2008). The Indian call center experience: A case study in changing discourses of identity, identification, and career in a global context. *Journal of Business Communication, 45*(31), 32–60.

Paulet, R. (2008). Location matters: The impact of place on call centres. *Journal of Industrial Relations, 50*(2), 305–18. http://dx.doi.org/10.1177/0022185607087904

Peck, J. (1996). *Work place: The social regulations of labour markets*. New York: Guilford Press.

Perez, K. (2005). 1–800-(Re)Colonize: A feminist postcolonial and performance analysis of call center agents in India performing U.S. cultural identity. *Electronic Theses and Dissertations*, paper 279. Retrieved March 2013 from http://digitalcommons.library.umaine.edu/etd/279.

Poster, W. (2007). Who's on the line? Indian call center agents pose as Americans for U.S. outsourced firms. *Industrial Relations, 46*(2), 271–304. Retrieved 7 January 2015 from http://dx.doi.org/10.1111/j.1468-232X.2007.00468.x

Ramakrishnan, S. (2011, 11 March). Contact center industry: Where do you go from here. *Cyber Media*. Retrieved March 2013 https://issuu.com/globalservicesmedia/docs/march2011.

Shome, R. (2003). Space matters. The power and practice of space. *Communication Theory, 13*(1), 39–56. http://dx.doi.org/10.1111/j.1468-2885.2003.tb00281.x

Shome, R. (2006). Thinking through the diaspora: Call centers, India, and a new politics or hybridity. *International Journal of Cultural Studies, 9*(1), 105–24. http://dx.doi.org/10.1177/1367877906061167

Sinibali, C. (2008, 8 July). Lampedusa's migrants. *Guardian.com*. Retrieved 25 March 2014 from http://www.theguardian.com/world/gallery/2008/jul/04/2

Skey, M. (2011). *National belonging and everyday life: The significance of nationhood in an uncertain world*. New York: Palgrave Macmillan http://dx.doi.org/10.1057/9780230353893

Smolin, J. (2011). Burning the past: Moroccan cinema of illegal immigration. *South Central Review, 28*(1), 74–89. http://dx.doi.org/10.1353/scr.2011.0011

Sonntag, S. (2005). Appropriating identity or cultivating capital? Global English in offshoring service industries. *Anthropology of Work Review, 26*(1), 13–20. http://dx.doi.org/10.1525/awr.2005.26.1.13

Taylor, P., & Bain, P. (2005). India calling to the far away towns: The call centre labour process and globalization. *Work, Employment & Society, 19*(2), 261–82. http://dx.doi.org/10.1177/0950017005053170

Taylor, G., & Van Every, E. (1999). *The emergent organization: Communication as its site and surface*. Mahwah, NJ: Lawrence Erlbaum Associates http://dx.doi.org/10.1177/0950017005053170

Thompson, P. (2003). Fantasy Island: A labor process critique of the age of surveillance. *Surveillance & Society, 1*(2), 138–51.

Tracy, S., & Trethewey, A. (2005). Fracturing the real-self fake-self dichotomy: Moving toward crystallized organizational discourses and identities. *Communication Theory, 15*(2), 168–95.

Tsagarousianou, R. (2004). Rethinking the concept of diaspora: Mobility, connectivity and communication in a globalised world. *Westminster Papers in Communication and Culture, 1*(1), 52–65.

Verheecke, L. (2014, 11 April). Morocco's call center workers get organized. *Equal Times*. Retrieved 10 November 2015 from http://www.equaltimes.org/moroccos-call-centre-workers-get-organised?lang=en#.VkObM02FOKR

World Bank Report. (2012, 18 October). Retrieved 30 October 2015 from http://data.worldbank.org/indicator/SL.UEM.1524.ZS

Wortabet, J. (1995). *Hippocrene standard dictionary: Arabic-English, English Arabic*. New York: Hippocrene Books.

Yabladi.com. (2013, 27 March). *Immigration: Le Maroc accueille des travailleurs pauvres espagnols*. Retrieved 14 April 2016 from http://www.yabiladi.com/articles/details/16304/immigration-maroc-accueille-travailleurs-pauvres.html

8 Border Speech between Two National Linguistic Ideologies: The Case of Bilingual El Paso Call Centres

JOSIAH HEYMAN AND AMADO ALARCÓN

Introduction

This chapter examines the contribution of new production processes and markets in the information economy to the redefinition of national linguistic ideologies in the context of the U.S.-Mexican border, specifically in El Paso, Texas. We focus on the call centre sector – an industry whose basic work material is language and whose product is conversations – in El Paso, a bicultural border locale with a Mexican-origin population majority, linguistically diglossic[1] in English and Spanish, with a variety of individual repertoires in the two languages.

The association between fluency in Spanish, limited proficiency in English, and work roles with a low communicative profile and a low position in the work hierarchy has been a historically deep and persistent trait of the socio-labour structure of El Paso. Before the North American Free Trade Agreement (NAFTA) came into effect in 1994, the main industry in El Paso was the manufacture of men's slacks. This industry was characterized by a poorly paid labour force of Mexican origin and Taylorist-Fordist work regimes; the operatives were mainly immigrant women or commuters from Mexico (Spener, 2002), and their linguistic skills (little or no competence in English, and low levels of schooling) were largely irrelevant to the demands of production (consistent with analyses in Boutet, 2001, p. 56, and Cohen, 2009, p. 26). Interestingly, this pattern became established on the Mexican side of the border from 1965 with the maquiladora export-oriented assembly industry, which was reinforced by capital mobility after NAFTA came into effect, whose relaxed legal restrictions on the garment assembly industry led to a rapid exit from El Paso.

With the rise of the information economy and society (Castells, 1996) corporations have become new, powerful agents that, through corporate and linguistic ideologies, enact language management, modify linguistic practices, and influence public beliefs (Spolsky, 2009). In the case of El Paso, this activity is represented by the call centre industry, much of which is bilingual and thus centred on the management of two languages and the interactions between them. With this in mind, our objective is to learn if and how corporations in the new service economy, especially call centres, are modifying identities and linguistic ideologies in diglossic and border contexts.

Hypothetically, the emergence of bilingual call centres on the border, demanding of employees a competence in both Spanish and English, should contribute at the "micro" level to the reversal of the previous patterns of linguistic and cultural change (from Spanish to English) through the treatment of Spanish as an element of human capital that is valued in the market (Breton, 1998; Grenier, 1982; Grin, 1999). Furthermore, the cultural richness of borders, in particular Anglo-Hispanic biculturalism, ought to contribute to a better positioning of corporations in the global service economy and, by extension, provide better opportunities for bilingual and bicultural workers. These processes, in turn, should contribute at the "macro" level to linguistic ideologies of a revitalizing character, in which the Spanish language would constitute a space of affirmation and resistance of the Hispanic community (Lago, 2009, p. 24). However, these optimistic assertions have not been confirmed by our research in the economic sectors such as health, translation and interpretation, cleaning, and call centres, where we have analysed organizational structures and their connection with the use and valuation of linguistic competencies in Spanish and English (Alarcón & Heyman, 2012, 2013, 2014).

El Paso call centres are predominantly composed of bilingual Mexican American and Mexican employees, with a considerable overrepresentation of Anglo-American managers. Their market is mainly the English- and Spanish-speaking populations of the United States, although calls in Spanish only constitute between 2 and 8 per cent of the total, depending on the work unit analysed (Alarcón & Heyman, 2013). It is difficult to justify the orientation towards the internal (national) market as a paradigmatic example of the new global economy (Villa & Villa, 2005, pp. 172–3). Rather, we examine El Paso call centres as an instance of the effects of international linguistic and social borders on an industry (contained within national boundaries) based on intensive use of language.

We find that the ideologies underlying the politics and practices of linguistic management of bilingual call centres do not contribute to a greater valuation of Spanish as a technical competence meriting formalization and reward. Rather, national linguistic ideologies, mainly from the United States but also Mexico, retain their traditional weight even when they are operating in new contexts in corporate structures and ideologies (Boix & Vila, 1998, p. 159). This involves reproducing Spanish on the U.S. side of the border, to be viewed and managed as a heritage language,[2] an authentic social competency (Heller, 2003, 2010), which is subordinate to English (considered the language of corporate power, prestige, and reward) and also indirectly subordinate, by negative comparison, to prestige forms of national Spanish within Mexico. We see that the large U.S. corporations in El Paso choose to manage Spanish in call centres as if it were a "natural" attribute of the local setting, generally not worthy of reward.

This managerial conceptualization of Spanish is consistent, on the one hand, with local ideologies (Achugar & Pessoa, 2009; Timmons, 1990; Velázquez, 2009) and, on the other hand, with neo-Taylorist management practices common in mass-market call centres, which combine (1) standardization (Cameron, 2000, p. 224; Milroy & Milroy, 1998), especially through scripts and intensive control of conversations (Leidner, 1993; Poster, 2007); (2) turnover of personnel and a low need for investment in training (Del Bono, 2001); and (3) permissiveness with local accents as a means of authentic relations with immigrant clients from related source countries and language registers[3] (Heller, 2003, 2010). These managerial strategies permit access to abundant labour at low wages, especially in the case of Spanish-speaking workers in the United States, given that Spanish is a language with low social prestige there. As a result the added cost of Spanish to these organizations is almost nil. By being caught in a border position between two national-linguistic monoliths, border bilinguals are devalued in the emerging communication service economy.

Ideologies, Borders, and Call Centres

We can understand linguistic ideology to be a combination of the beliefs about language on the part of speakers, of that language and others, that rationalize and justify the perception of linguistic structure and practices (Silverstein, 1979, p. 193). Linguistic ideologies, conscious and unconscious, are the indirect, non-mechanistic result of positions, interests, and social practices of individuals and groups (Boix & Vila, 1998; Schieffelin & Woolard, 1994). In situations of linguistic diversity, diglossic ideologies (Boyer, 2006) combine beliefs about unequal structures and roles of languages and the unequal distribution of power among social and linguistic groups.

Within the major social transformations of recent decades, linguistic ideologies are being reworked and re-legitimated. The linguistic ideologies that legitimized homogeneity on the basis of modernizing or imperialistic arguments now need arguments related to "democratic access to a shared market that can respond better to the specific needs and interests of Spanish-speaking producers and consumers than can its English-speaking global competitors" (Heller, 2010, p. 106). Nonetheless, the legacy of linguistic systems of large states continues to make possible a privileged control of resources (economic, legal, cultural, and so on) on the part of socially and linguistically dominant groups. In accord with Heller, "the work that needs to be done in the neo-neo-colonial project is thus focused less on the language teaching and translation that was the hallmark of earlier forms of empire (although those remain) than on relegitimizing those activities and constructing new subjectivities. Equally important and difficult is the work of managing the debates about what counts as legitimate English (or Spanish, or French, or Portuguese) and who defines it – debates opened up by re-framing colonial relations on a supposedly more equitable basis" (Heller, 2010, p.106; Gal & Woolard, 2001; Urciuoli, 2008).

The persistence of old linguistic structures and their re-legitimization is made especially evident in intergenerational transmission. In agreement with Boyer (2006, p. 4), in diglossic contexts "it is not the children who stop learning the language of the parents, but the parents that, most often victims of guilt …, of self-deprecation, products of a diglossic ideology… no longer pass on the dominated language to their children." The process of the hierarchy reproduction and linguistic ideologies applied to the new working class of the call centres enables us to understand the lack of interest in formal Spanish education that persists among many Latin@s[4] in the United States; it is observable in the limited and informal intergenerational transmission practices of Mexican Americans in El Paso, in spite of the existence of a greater vitality of Spanish in public spaces, including the marketplace, as well as more favourable attitudes towards the language among the general public (Callahan, Wilkinson, & Muller, 2010; Velázquez, 2009).

Language management policies of corporations in the contemporary global economy (and, relatedly, the service economy) face important contradictions (Duchêne & Heller, 2011; Grin, Sfreddo, & Vaillancourt, 2010; Kelly-Holmes & Mautner, 2010; Tan & Rubdy, 2008; Urciuoli & LaDousa, 2013). Policy implementation is carried out in older ideological frames that presume, consistent with the national frame of the region of our study (Hudson, Cháves, & Bills, 1995; Veltman, 1988), and reproduce the subordination of Spanish in the context of the contemporary economy. These contradictions demonstrate within globalization the emerging relations between the beliefs and behaviours of speakers and the wider economic, social, and cultural structures.

Call centre workers, much like border society as a whole, are caught in a multi-positioned web of conflict over U.S. and Mexican nationalism (Vila, 2000).[5] There is a range of stances, including strident nationalisms of both sides and various nuanced identities in between, as well as an anti-nationalist (to some extent) identity as a "borderlander." This has direct expression in complicated and tense linguistic ideologies. English – specifically the formal national standard – is often understood as a superior language, associated with economic modernity and prosperity, and filled with U.S. assimilationism and patriotism. As Ofelia García and Leah Mason (2009, p. 89) write, "by constructing Spanish as a language of poverty and Spanish-English bilingualism as nonexistent, White English monolinguals enjoy privilege while excluding US Latinos." This U.S. English ideology is subtly but pervasively stigmatizing towards fluent bilingual borderlanders who have either a Spanish-language accent or the "Chicano English" version of native-speaker English, though regional responses ignore or resist these negative categorizations (Martínez, 2006). Chicano English is a set of accents that resemble neighbouring U.S. dialects (southern and western) but have modest Spanish-influenced prosody[6] and pronunciation, some words and catchphrases, and occasionally syntactic borrowings.[7] In the region it is rarely ideologically marked and, if noticed, has low prestige.

More visible and conflicted is a cluster of linguistic patterns from the mixture of the two languages, colloquially called Spanglish. Spanglish can include code switching[8] between the two languages, and recent English borrow words in Spanish (and an older substrate of Spanish borrow words in English), and there are substantial changes in syntax, prosody, and pronunciation going in both directions. Spanglish is ideologically highly disputed. It is devalued from both English and Spanish purist stances but viewed by its own speakers with feelings of cleverness (wordplay), intimacy, and solidarity of the working class. However, it almost never has prestige in the two wider societies. As a result,

English patterns in the El Paso border region involve speech register ranking, with a mid-Atlantic-based, white, national English standard ranked above regional Chicano English, Spanish, and mixes of the two.

These linguistic ideologies emerged historically with the U.S. military and the subsequent legal and economic conquest of former Mexican territories, such as El Paso (Montejano, 1987). An important pattern beginning in the 1880s and continuing to the present is the association of Mexican origin and Spanish language with subordinate, low-wage labour; language is part of the racialization of Mexican-origin people and, more widely, "Hispanics" in the United States. Immigration-driven social change, superficial political correctness, and enduring civil rights struggles by Mexican-origin people have tempered these ideologies to some extent, engendering contradictory positive and negative valuations of Spanish in the public sphere (García & Mason 2009). Nevertheless, the border region especially continues to be stigmatized in the rest of the United States, through which the U.S. nation is in part defined by a contrast with the perils emanating from Mexico (Ganster & Lorey, 2008, p. xxi; Heyman, 2012). Just as the border as a symbol condenses stigmatizing ideologies about Mexico within the United States, in Mexico the border condenses defensive national ideologies about the United States, often involving language.

Mexican linguistic ideology about Spanish is marked, with a particular emphasis on the preservation of Spanish. This is a result of Mexico's protest nationalism against the domineering power of the north, which had once conquered and stolen half the national territory and currently mistreats immigrants. At the same time, educated Mexicans historically have been very hostile to Mexican emigrants to the United States for betraying the homeland, although this is changing (the reality, of course, is that emigrants often leave with few choices under desperate circumstances). Relatedly, nationalists in Mexico have been suspicious of the northern border as a region in which national culture and identity is eroded by the powerful proximity of the United States. In our research we mainly found manifestations of Spanish linguistic purism centred on a national educational standard, which serves to resist the Americanization of Spanish (for example, Spanglish). Prestige-register Spanish that is based on school training is also a performance of class superiority because university-educated Mexicans can perform this standard, unlike the many Mexican Americans who speak the uneducated peasant-worker registers of their emigrant origins.[9] In the U.S. educational context many Mexican Americans have little or no schooling in Spanish, are not literate in it, and learn Spanish via interpersonal interactions (for example, with older relatives). Their oral code differs considerably from Mexican codes shaped by written standards. We thus refer to their Spanish as a heritage language, acquired within the community with little or no formal training. Their heritage Spanish is separated from and ranked below both educated U.S. English and educated Mexican Spanish. However, such accents, vocabularies, and registers may seem welcoming and unpretentious to the working-class Latin American immigrants who are the main clientele for Spanish-language calls in call centres. The typical border linguistic performances are thus caught in a largely negative space between two nationalistic language ideologies. In situations of comparison between the two dominant codes, moreover, English outranks Spanish. The border forms, however, are sometimes positively valued, mainly in interactions between speakers of equivalent low prestige.

To help us understand the way in which this field of contending regional linguistic ideologies fits into the new communicative economy, we begin with a key observation by Heller (2003, 2010). The linguistic policies of corporations in the contemporary economy move back and forth between the search for Taylorist standardization of language in providing services – demonstrated by call centre scripts as a system of control over "speech and talk" – and the value of the authenticity of such talk in a context of saturation of products and increasing competition. The distinction achieved by authenticity and affective-identificatory bonds with clients can be key to business success. In the border setting this means facing alternatives between formal codes in each nationalized language and the more "authentic" borderlander speech patterns. Likewise, according to Heller, the businesses oscillate between viewing language as a technical competency, and as such directly compensable, or as a social competency and innate talent of people, making invisible their abilities and rendering unnecessary the compensation of and investment in these competencies. Again, the former corresponds more to nationalistic linguistic ideologies, and the latter to heritage Spanish and English, and Spanglish, of the U.S. borderlands.

A good part of the research on language and call centres has focused on interactions across nations within the domain of world Englishes.[10] Many, though not all, of the studies address call centres in South Asia that serve an Anglo-Saxon clientele (Morgan & Ramanathan, 2009; Poster, 2007; Rahman, 2009; Taylor & Bain, 2005). The relationship is often understood and performed through the frame of nationalistic language ideologies and linguistic actions in complex interplay among workers, managers, and clients (Cowie, 2007; Mirchandani, 2004, 2005; Rahman, 2009; Sonntag, 2009; Taylor & Bain, 2005). There is a modest though valuable literature on call centres within multilingual nations, though involving pre-existing national linguistic groups rather than new immigrants (for example, Roy, 2003). Our work, however, addresses a complex situation brought about by migration, both for the labour force and for the target public. It concerns call centre conversations within one nation (the United States), though it is affected by another nation (Mexico). There are two languages involved, English and Spanish (and even mixtures of the two), and this brings in language ideologies of two nations with different though related histories and patterns of social inequality. Owing to the dense interplay of the two languages and countries, our account adds subtlety to a call centre literature that treats language accents and ideologies, with some exceptions (for example, Cowie, 2007), in fairly schematic, either-or ways (for example, local operators versus distant consumers). In our study the internal tensions among and between workers and managers prove to be as important as the operator-client interactions. The extensive study of Spanish and Spanish-English bilingualism in the border region and the entire United States provides a background on sociolinguistic hierarchies that strengthens the interpretation of such language struggles in call centres, rendering El Paso a valuable site for a case study. We thus bring to the call centre literature new questions about immigration, emergent multilingual societies, and the relationship of international nationalisms to internal sociolinguistic processes.

A Border Location in the New Economy: El Paso

The bilingual call centres of El Paso thrive as much from shared social representations about language as a mechanism of legitimization of corporate management, as from the different and specific levels of individual bilingualism available in the market. In El Paso there has been, on the one hand, a notable situation of diglossia, with Spanish as a subordinate language and English as the prestige language in public life and institutions. This situation originated in the economic and political control that the Anglo-American (an ethno-racial term notable for using language, English, for its definition) minority has exercised in the region (Timmons, 1990). On the other hand, with respect to individual bilingualism, the American Community Survey in 2008 found that 73.9 per cent of the

El Paso population spoke Spanish at home, and 70.7 per cent spoke English very well. Almost all Mexicans (people born in that country, as opposed to Mexican Americans), who were 24.3 per cent of the population, maintained Spanish at home (97.3 per cent), but only 30.4 per cent indicated that they had very good knowledge of English. Among Mexican Americans (52.3 per cent of the population), 84.6 per cent used Spanish in the home, and 82.9 per cent indicated that they had very good knowledge of English. Among non-Hispanic whites (locally, Anglos), who were 12.9 per cent of the population, 12.5 per cent reported using Spanish in the home. Of Spanish speakers in El Paso, 28.9 per cent were below the poverty line, as opposed to 14.1 per cent of those who spoke English only (American Community Survey, 2008). Bilingualism indices imply important differences in the prestige of activities and occupational sectors (Gibson, Olmedo, & Caire, 2008).

As the principal local newspaper optimistically proclaims, El Paso has turned into "a critical part of the call center–service center ecosystem" that "has grown to be one of the city's largest private employers, has continued to hire people, even during the recession," all while the sector stagnates in the United States (Kolenc, 2010). The first bilingual call centres were established in El Paso by the end of the 1980s, but it was only by the late 1990s to the early 2000s that large corporations committed themselves to the city. According to the Quarterly Census of Employment and Wages, telephone call centers (NAICS 6056142) employed 1,434 in the private sector in 2001 (not including workers in emergency dispatch and so on), and the number had grown to 7,338 workers by 2009. If we adopt a wider definition of the sector that includes thirty-seven different economic activities involving work that specializes in telephonic contact with clients (Moss et al., 2008; Wirtz, 2001), call centres in El Paso employ 13,992 workers in thirty-four work centres (REDCO, 2010).

These thirty-four call centres receive calls and offer mainly inbound customer service. Outbound service (initiating calls), such as conducting questionnaires for market research and analysis, is the main activity for a minority of these centres. With respect to ownership, there are more in-house (directly corporate-owned centres) than outsourced sites, but the former are more limited in numbers of positions (workstations) and employees. The centres are mainly large U.S. corporations that almost exclusively service the U.S. market. The El Paso call centres primarily service mass markets. There are only a few cases in which there are small centres or their market can be considered a "professional service model" typical of business-to-business centres (on types of centres, see Batt, 2002; Heskett, Sasser, & Schlesinger, 1997).

Following the weighted data derived from the Public Use Microdata Sample (PUMS) 2006–8, concerning customer service representatives and telemarketers (n = 292), the majority of workers are women (60.3 per cent). Concerning their educational level, 39.7 per cent have no college education, 46.2 per cent have some college or more, and 14.0 per cent have a bachelor's degree or higher. The average age is 29.98, with a standard deviation of 11.41.[11] The largest group, 64.9 per cent, are Mexican Americans; the second group, 16.2 per cent, are Mexicans; and in third place, 8.3 per cent, are non-Hispanic whites, even though they are over-represented in managerial and intermediary positions (as we observed in fieldwork). The remaining 10.6 per cent belong to other groups, including 4.8 per cent who are non-Mexican Hispanics. Of these customer service representatives and telemarketers, 79.0 per cent report speaking Spanish at home, and 11.6 per cent report speaking English "less than very well." The demographics of these workers, that they are mainly Mexican American, with relatively limited education and high degree of bilingual fluency, with a smaller group of workers having limited English (possibly Mexican immigrants), is the key to our research.

Our interviews, conducted in 2010, sought to represent the full production hierarchy in call centres.[12] We spoke to operators, trainers, managers, and language professionals (interpreters and translators). Operators are numerically most common, but to capture this hierarchy we over-represented the latter three categories. Of the thirty-one operators interviewed, nineteen were women, and twelve were men. Four were immigrants from Mexico, and the rest Mexican Americans. Of the twelve language professionals interviewed (substantially over-represented, following our research focus on language and the new information economy), nine had been educated in Mexico, and three in the United States; seven were women. We interviewed eight executives and intermediate managers – four men and four women. All six intermediate managers interviewed were bilingual, of Mexican origin, and locally enculturated. As for the two executives, both had been assigned to El Paso from outside the border region, though one was originally from El Paso, of Mexican American origin; the other was Anglo-American (the local term for "white"). Both were bilingual.

An important element of ambiguity in the borderlands is that some middle- and upper-class Mexicans give birth to children in the United

States, who then have U.S. citizenship. Often they are educated in whole or in part in Mexico and return to the United States for university and careers. They are not counted as immigrants (foreign born), but linguistically and socially they are Mexican. In our interviews they were found extensively to be in language professions, training, and management, areas of work that take advantage of their educational and class backgrounds. Whatever their legal status (which is stated in a mechanical fashion in the PUMS data), in call centres they operate culturally and linguistically as national Mexicans, which proved important in our analysis.

A large part of the business-press literature on call centres focuses on questions of location. Call centres are characterized as being very mobile except for some legal restrictions (for example, on certain insurance contracts) and culture-related security risks that affect corporate offshoring decisions. Aside from these restrictions, locations are chosen according to factors of production, above all the supply of workers with the appropriate linguistic abilities, and their wage price. What does El Paso offer in order to attract and retain call centres in the face of offshoring to other Latin American destinations that have a supply of bilingual labour, available for less pay (Rojas, 2007)? For example, some Latin American call centres now employ English-dominant or bilingual deported Americans in Mexico, El Salvador, and so on, as an offshore strategy for the U.S. market. In general, call centres are strongly affected by the linguistic availability of sites within an international division of labour, which often "involve[s] connecting service-seeking customers from wealthy, English-speaking countries with service-providing workers from more multilingual, less economically powerful states or regions" (King, 2009, p. 1). Thus, corporate localization policies can demand as much *intensive control over national identity* in contexts where salaries are globally low but locally high (for example, the case of north India studied by Poster, 2007, p. 293), as they demand *authenticity of speech* in contexts where local wages are relative low by national standards but globally high (for example, the case of Switzerland analysed by Duchêne, 2009, p. 48), to cite two extremes.

In the case of El Paso, according to representatives of REDCO, the principal local consulting organization dedicated to obtaining investment for the El Paso area, which is in continuous consultation with the call centres, there is a combination of reasons that call centres invest in El Paso: (a) a bilingual labour force, (b) low wages and cost of living, and (c) no need to pay extra for the bilingual workforce.[13] The analysis of statistical sources allows us to generalize this situation along the border and in other locations with a Hispanic majority in the United States.

Average pay, as cited in the Occupational Employment Statistics (OES) for May 2009, is $11.23 per hour for customer service representatives and $9.30 for telemarketers, placing El Paso among the lowest in the country (at number 543 of the 577 OES geographic units). In the thirty-four locations with the worst pay, a number of sites stand out distinctly as "bilingual locations" in linguistic and wage competition with El Paso. The Puerto Rico Non-metropolitan Area 2 (east Puerto Rico), at $7.76 per hour, has the lowest pay rate. We also find Ponce, Puerto Rico, to be below El Paso at $9.60, and San Juan-Caguas-Guaynabo, Puerto Rico, with an average wage that is $0.20 higher than that in El Paso. On the border with Mexico, McAllen-Edinburg-Mission, Texas – the MSA that has the highest percentage of Spanish speakers in the fifty U.S. states (slightly higher than El Paso's) – averages $9.72 per hour. Near the border, Corpus Christi, Texas, has 4,668 workers with an average salary of $10.80 per hour.

There is no indication that the large concentrations of Hispanic workers with their linguistic "capital" in Puerto Rico or the Mexican-U.S. border favour better salaries than do the mainly monolingual call centre sites in the rest of the United States, even though this linguistic richness certainly acts positively to attract investment. As a local worker of Mexican origin, who had risen into management of training programs, told us, "this call centre is here because of cheap labour. There are [in this corporation] two other centres: one in Illinois and one in Florida. And they pay much more ... partly because of the cost of living. Basically, they have a different perception of us, the workforce here, than they have of those in Illinois [the corporate headquarters site]."

This overview leads us to propose that the border in this industry, as with the assembly plant industry in El Paso's past (and Mexico's present), serves to attract "cheap work" investment in the new economy (Alarcón & Heyman, 2013, 2014).

Competing Ideologies: Representations of Language in the New Economy on the Border

Concerning the linguistic characteristics of El Paso, a Mexican American local economic-promotion officer described the views of central corporate offices about languages in El Paso as fitting into the framework of "negative feelings about the border community." El Paso's location, on the border between the United States and Mexico, awakens negative sentiments towards Mexico, long the target of negative stereotyping in the United States. This is exacerbated by El Paso's proximity to Ciudad Juárez, Mexico, a city stigmatized by violence. El Paso also suffers from internal U.S. stereotypes, owing to racism towards its majority Hispanic population, as well as its high poverty rates and its relatively large geographic distance from various corporate headquarters. Taken together, these factors often lead to El Paso being considered "on shore" in legal terms but "near shore" in some ways by executives in terms of cultural and geographic distance from home offices. The informant said that "the managers [Anglo Americans] in home offices think El Pasoans do not speak English. They are preoccupied with the idea that El Pasoans cannot give quality service in English and of course assume their level of Spanish is sufficient, or don't even consider this question." One upper corporate manager stated in a trade journal: "There's minimal difference in treating someone who speaks Spanish if they are from Mexico living in Texas or if they are from Cuba or the Dominican Republic living in south Florida ... It's no different than treating a Texan or a New Yorker or a Canadian. As long as you give them good service and sell them what they want, that's what matters (Read, 2002)."

In a few corporations, especially ones with legal responsibilities (for example, insurance), there are some tendencies towards certification of Spanish language skills for official interpretation and translation duties, which benefit educated Mexican immigrants (discussed below). For most workers, however, the skills used by the corporation are either purely English – for which being a border Hispanic is a sign of exploitability (Alarcón & Heyman, 2013) without real concern about language – or flexibly bilingual through code switching with clients who need that. But when there is no distinction on the job about when Spanish needs to be used, and it occurs spontaneously, it is difficult or impossible to reward or cultivate it.

There is, in most cases, hardly any linguistic selectivity in either language at initial recruitment (which usually involves just reading an oral speech test, fundamental to script-based production in call centres). There is also very little investment in linguistic training or formalization that could result in paying extra premiums. Both workers and managers widely reported this tendency, including two workers who reported the centre's initial – but rudimentary – concern with their ability to read scripts in the two languages, and no concern with identifying skill levels beyond that: "'Now read this (in English). Now read this in Spanish.' To know if you can read. But very informal, everything" (Operator). "They made me read one page in English and another in Spanish, to make sure I could read the two languages well. So, if you can read it, you can speak it" (Operator).

Despite corporate concerns with English proficiency, it receives little selectivity or investment. The assumption is that local educational institutions will produce a satisfactory level for the corporations. The tendency to under-invest in linguistic training facilitates exploitation and is even more pronounced for Spanish compared to English. As informants repeatedly emphasized, corporate management had little concept of Spanish. As one trainer said, the attitude is: "If they [El Pasoans] understand Spanish, then they can sell in it." A local economic-promotion agent who deals with many such corporations said, "They [the central managers] do not have a very good idea about what is Spanish." English skill levels, accents, etc., are visible to central corporate staff, but comparable aspects of Spanish are unmarked. This fits a wider identification of the U.S. nation state with English and invisibility or symbolic but superficial visibility of Spanish (García & Mason, 2009).

This also holds for training in either language, but especially Spanish. Training periods range from one to fifteen days, although by the second or third day the focus shifts to making sales and attending to clients. Training varies considerably between call centres, from the simple practice of rote reading of scripts to clients at the start of employment, to extended and substantial programs of continuous training. Nevertheless, the in-depth training programs do little to address language as such, and the linguistic training that exists follows a uniform national standard, which addresses only English and not Spanish. An in-depth training manual from a large and high-quality call centre included, in approximately equal parts, information on the product and business, call centre information systems, and communication processes in national standard English (for example, empathy and politeness). In other words, there was little attention to the tasks and skills of bilingual workers. Only at one large and prestigious company were there a few pages in the training manual addressing Spanish-language issues, such as polite forms to open and close conversations. As a training director told us, "in my folder of four hundred pages I have two pages of etiquette in the Spanish language. It is the last thing. Nothing more than an exercise where you ask how to say 'engine' in English. Please! They should put in a little etiquette! That you should listen to the client! This is a training manual!" It is interesting that communication processes are in most cases viewed at the corporate level as uninfluenced by the language used and that English is the unmarked standard for such training (hence, the training director's complaint that the Spanish language section provided only a brief exercise in asking questions and did not broach in Spanish the metalinguistic task of effectively listening to clients).

Ironically, there is greater concern about English than Spanish, though not enough to discourage hiring low-wage border workers. The negative evaluations about linguistic competency in English focus on the southern accent, Mexican American accent, lower-class register, excessive use of code switching between English and Spanish (one form of Spanglish), and uneven ability to convert reading into speaking (literacy problems). Interestingly, following the 2006–8 PUMS, 11.6 per cent of operators in this sector report speaking English less than "very well" (of course, implying that most workers do speak English very well). They are not being hired specifically for Spanish skills, we think, since in our fieldwork we did not observe workstations exclusively serving Spanish speakers. Rather, this hiring seems to be the outer edge of a larger pattern of hiring cheap and vulnerable labour, whether it be fluently bilingual or limitedly proficient in English.

Historically, limited English proficiency in the borderlands has been a key marker of cheap and vulnerable labour in the border region (Alarcón & Heyman, 2014). Employers often assume that such workers can gain sufficient linguistic competency through practice (learning by doing). One training director said:

> Last week there were six people in the class. They all came from Mexico, and they all had problems speaking English. They understood almost nothing. Because I told them to go to page six, and I went to the back of the room to see in the computers if they were doing what I told them, and they did not know what was happening. And they were always asking their neighbour. Many of them, because of embarrassment, said that everything was fine, they were going to pay more attention. But one said, "It is that I don't know English well [*laughs*]." And then I said to him, "OK, it's fine, I am going to put the instructions on the blackboard so you can see the page number," and then they were able to follow. They followed in English and Spanish. In both.

This interview, in part, reflected the situation during our fieldwork when call centres were rapidly hiring people who had fled to the United

States from violence in Ciudad Juárez and who often had limited English skills but were distinctly needy. This recruitment and training strategy is consistent with the objective of minimizing labour costs in the mass-market service model, at the expense of the linguistic quality of service (Alarcón & Heyman, 2013).

El Paso Mexican Americans generally perceive a stigmatization of their English-language speech by people from the U.S. interior, including central offices. In this discussion we use a social-geographically meaningful distinction between the region on or near the U.S.-Mexican border, an area with extensive knowledge of and interaction with Mexico, versus the remainder of the country where most people do not have this knowledge and interaction. The response is the deployment of an ideology of positive localism: the so-called Chicano English and the commonplace bilingualism of the El Paso population show that "everyone has an accent in English." This insistence on the legitimacy of the local accent certainly resists corporate home office stigmatization.

In this regard, Mexican American middle management, fluent bilingual workers (whose Spanish is usually informal), and those workers who are less fluent in either English or Spanish, the latter two in continual contact with clients, argue that language-related complaints contain more a dissatisfaction with the service offered by the business than linguistic limitations. This came out in a nuanced interview that considered both cases of accent conflict and the dismissal of other such complaints to be actually products of the business process:

> To New Yoricans [Puerto Ricans from New York], African Americans, at times they are not understood … Sometimes they [clients] say to me that they want to be transferred to someone who speaks English. But in general, yes, I speak well in English. I do have workmates who have a very strong accent. Then, clients don't want to speak to them. But sometimes it is an excuse by the client, a way to say that they do not want to purchase some service. If the agent does not do her job well, then if the client is focusing on that, it is a problem of communication. But really, this is a problem of company policy, because the client wants to have it all. (Customer service representative)

In terms of Spanish, Anglo-American managers and Mexican American operators share a view of Spanish in communication with a Hispanic public – as a heritage language, with a "high degree of variability in the personal, cultural, socio-demographic and linguistic traits deemed central to the meaning of the term" (Wiley, 2001). An illustrative phrasing is that the population is "bilingual fluent in Spanish." This characterization by local promoters of the population as "bilingual fluent in Spanish" emphasizes to the investors the Americanness of El Pasoans and that as Americans they are English speakers *with the addition of a heritage language*. In no case are they viewed as a population linguistically educated in Spanish. Within this understanding, competency in Spanish is a socially abundant resource in El Paso, and considered sufficient to attend to the Hispanic public of the United States. While not all Hispanic telephonic clients are of Mexican origin (about 60 per cent of current U.S. Hispanics are), they tend to share with El Paso operators similar working-class registers in Spanish, selected by the in-migration process, as well as fluid English influences on Spanish. Hence, even if their accents and dialects are dissimilar (for example, between a Mexican and a Peruvian), their shared informality provides a feeling of informality and solidarity that is valuable to the production process and would not occur with an educated standard.

The relatively few employees who hold positions as language professionals in call centres are young people with advanced education, the majority of it obtained in Mexico, and some with a U.S. university degree. Their tasks in the various call centres range from translation of scripts, recruitment (including examination of linguistic ability), training, and work in small departments of interpretation and translation (for example, translation into English of conversations between clients and operators with legal implications). They consider themselves to be the only legitimate speakers of Spanish in the area, distinguishing themselves from the linguistic poverty of the region, which brings them to develop a very critical discourse about the role of Spanish in the corporations.

Mexicans in call centres who have a university education are critical of the Spanish competency of Mexican Americans, for reasons listed earlier in the chapter. Notably, university-educated Mexicans criticize the Spanish of Mexican Americans in ways that highlight the key issues in the capital-labour relation, specifically the supply and valuation of linguistic capabilities as a factor of production within the call centre industry. For example, a characteristic criticism they present is that "few [Mexican Americans] can pass a written entry exam" in Spanish, even though they can speak it. Another criticism is that there is "excessive" use of Spanglish, seen as a speech style associated with a "low cultural level" (ironically, among educated Mexicans, who usually learn English in school, scattering a bit of English in basically pure Spanish discourse is seen as highly prestigious). In these ways they interpret Spanish as a technical competency, acquired through effort in the educational system, and deserving extra compensation, rather than as a heritage language – by contrast with the view of Anglo-American managers and Mexican American workers. These language professionals assert their role at moments of ensuring the linguistic quality of service.

Ideally, these language professionals imagine a workforce that would employ an anonymous or neutral language for the entire Hispanic population, using scripts in Spanish. In practice these scripts reflect formalisms of the educated Spanish-speaking world, or at least Latin America. In some call centres, for example, code switching (Spanglish) is explicitly banned. There are likewise lists of prohibited words that derive from specific Spanish-speaking communities rather than cosmopolitan standards[14] (for example, for "tyre," *neumático* is proscribed over *llanta*[15] or *goma*[16] or the English borrow word *tyre*). However, this "verbal hygiene" practice remains far from being an integral stylization (Cameron, 2000, p. 326) of communicative practices. The total lack of control over the use of these listed words (in the few cases where such a proscription exists) points to an open conflict between Mexican Americans, grounded in a heritage-language repertoire, and the Mexican language professionals, as guardians of a "legitimate" code. Yet the language of the clientele is often closer to the heritage language of the post-immigrant operators – both in the technical features of their language and in the emotional tone of equal relationship – than with the educated code. In the following quote, a Mexican employee informs us that she also has ended up adapting to these informal practices "so as not to humiliate the client":

I want the person with whom I am talking to come away with the impression that I was a professional. And a professional does not speak Spanglish. I do not know where we have gotten this idea, because probably a professional does speak Spanglish. Now, let me tell you something. When you are an interpreter, let's say you are speaking with a client and you say to him, "Sr. Gonzalez, aquí tenemos su factura. – ¿Qué es eso? – El cobro que le mandan. – ¿Qué? – El bill." [Mr. Gonzalez, we have your bill here (with the proper Spanish word for *bill* used). What is that? The charge we sent. What? The bill (the English word in a Spanish utterance)]. Otherwise, it is not understood. "Estaba tanqueando cuando le dieron el golpe. ¿Y qué es tanquear? Pues llenar de combustible" [I was filling up the vehicle when it was hit. And what is *tanquear*? (the English noun *tank* for "gasoline," conjugated as a Spanish verb). To fill up with fuel.] Well then, you have to adapt to the needs of the clients so as not to humiliate them. (Training director)

The point about avoiding humiliation is interesting: it means abandoning insistence on a prestigious educated register, which would stand over working-class Hispanic immigrant clients, in order to use speech styles that are approximately equal to theirs. It brings into the U.S.-based call centres the sociolinguistic inequalities and conflicts of Spanish in Latin American countries, just as the central office versus borderlander tensions over regional English speech discussed earlier does for U.S. English hierarchies. And it conveys resistance based on worker-client similarities and commitment to service quality, in opposition to language ideology.

For Mexican-enculturated operators these national linguistic ideologies are reproduced. These operators felt entirely legitimate in their Spanish skills, identifying their Chicano colleagues as acculturated. With regard to those Mexican American operators, we observe an enormous variability in their capacity to communicate in Spanish in formal situations. Besides the effect of language shift from Spanish to English in the United States, most of them are descendants of working-class Mexican immigrants. However, they do not perceive their lower-register Mexican Spanish as bad. Despite these internal tensions among workers and supervisory staff, which reflect wider tensions in the borderlands (Vila, 2000), the overall framework set by U.S. capitalist corporations in the interior is little concerned with formal linguistic skills in English and almost not at all with formal skills in Spanish – in striking contrast to the ostensible agenda of the communication-intensive new economy – but rather is directed towards inexpensiveness and vulnerability of labour.

Hierarchies and Control

It is generally assumed in El Paso call centres that home corporate offices are made up of monolingual English-speaking "Anglos," and this is largely correct. It is widely resented and resisted. For example, the inclusion of Spanish as a work language within the organization creates important problems of control. Some Spanish-speaking supervisors and language professionals perceive an absolute absence of control over emails written in Spanish. "That's why we write emails in Spanish, so the superiors do not know what we are saying!" Yet this scenario is changing, albeit slowly and only at a few companies.

Call centres in El Paso are generally set up by Anglo-American executives and middle managers sent from the home corporate offices. This situation changes gradually with time because some of the managers return to their original sites after a period of service in El Paso. One reason for this is the intolerance of these Anglo-Americans, or at least some of them, of a local environment that is predominantly Hispanic, as El Paso has been in recent decades. Indeed, the city has undergone a marked out-migration of Anglo-Americans in recent years (Fernández, Howard, & Amastae, 2007). Another issue is that, from an organizational control perspective, if the operators are bilingual, their supervisors should be able to attend to complaints and incidents and monitor operations in Spanish. That implies that promotion should prioritize Spanish speakers. Locals generally fill vacancies such as assistant managers and technical staff, and the hierarchy gradually acquires a Hispanic character by organizational necessity.

> When the centre began, forty people came from Phoenix [a city in the interior], only Anglos. They created an organization, replicating the structure of Phoenix. But they also promoted people from the centre. Many people applied for promotion to supervisor. Now, they demand in all cases that supervisors be bilingual. The previous director who was an Anglo said that he wanted to return to Phoenix. He didn't speak Spanish. As for the associate directors, that previously were people from Phoenix, they demanded that they be bilingual since they have to struggle with supervisors and with all aspects of the centre. (Mexican American manager for a major telecommunications company, bilingual, born in the United States)

As with the control functions just discussed, demographic change with Latin American immigration throughout the interior of the United States also allows room for a relatively small group of Hispanics to advance in the company. There are two linguistic challenges: (1) mastery of interior American English standard, which particularly challenges educated Mexican immigrants whose first language is Spanish, but also challenges border-region Mexican Americans who have a Chicano English accent; and (2) mastery of high-register Spanish, which has corporate value (for example, in interpretation and translation) and benefits educated Mexican immigrants but devalues heritage Spanish speakers from the U.S. borderlands. This interweaves with significant social ruptures in moving to the Anglo-dominated interior.

In some of the major corporations, especially in-house call centres in companies with a high reputation and professional development plans, we observe a system of internal promotion of Hispanic talent. Through this, some Mexican and Mexican American workers (especially ones with higher educational levels) find opportunities for professional development. These promotion ladders frequently require time spent at the central corporate offices.

> We have something at work called a "talent profile." And you list your skills. It has to be something that is done by the company and you can provide the company ... "You know what, they want someone bilingual there." And sometimes they offer you work. To someone very well prepared for this. Let's say that they want two bilingual operators in Bloomington [deep in the U.S. interior], and you come from Mexico, and you are licensed to sell insurance (a state exam). You will start in a unit of operators that only focus on Spanish-speaking clients [versus the flexible, two-language, but English-dominant, operations in El Paso]. Because sometimes they need to call them from the central office to ask them something. You are not given many details ... I think this is only for Spanish-speaking clients. And over there, in the Great Lakes region, the Spanish-monolingual population is growing a lot.
>
> In that company, they have this culture, if you want to rise, you have to live in Bloomington for a while. You have to follow the same path as everyone. There you live in a very Anglo world. [*Question*: Did it scare you?] Yes, it scared me a lot. Here [El Paso] it is as if you were in Mexico, only with *pochos* (laughs) [a somewhat insulting word for Americanized Mexicans], with Spanglish. But there, I know that you would have very few [fellow] people. But we arranged to be with a group of purely Hispanics ... to make a barbeque [*laughs*]. We are from this group here in El Paso, a network within the company ... What they have done is promote Hispanic talent. (Large corporate training director, Mexican origin, university educated)

This informant is describing an imagined social context that differs from El Paso in important ways (see García & Mason, 2009). He has entered a corporate ladder in which Spanish-speaking "newcomers" are understood as commercially valuable isolates within the English-speaking nation. This contrasts with the fluid language switching and informal but effective cross-language translation seen in the border-lands. The way that he uses the linguistically and culturally insulting

Mexican term *pocho* for Americanized Mexicans indicates his conceptual distance from U.S. borderlanders as an educated Mexican immigrant. Yet at the same time he also identifies – as a label, a career path, and a corporate network – with the new ethnic identity of "Hispanic," which homogenizes and in some ways includes Latin American immigrants inside the United States. He thus leaves behind nationally ambiguous border identities such as "Mexican," typically used in El Paso as an ethnic term for U.S. people of Mexican origin.

Interestingly, this confronts him with U.S. linguistic nationalism that is partly avoided in the flexible setting of El Paso, placed as it is in a loosely controlled but highly exploited "near shore" production zone.

> Sometimes I am very worried about my accent. When I did my interview with the corporate office, I tried not to distinguish my accent. It is a problem ... They are Anglo-Saxons [a common usage in Spanish] at the level of management, and all that. If it were a Hispanic company, I could speak the way I do. But an Anglo-American business like this, in a position involving public communication, you have to speak without an accent. In the interview I just finished having, we discussed having clients from the north. It was obvious that they did not want anyone with an accent, but they cannot say that. It is a very notable situation. (Manager of interpretation and translation in a major corporation, bilingual, originally from Mexico)

Of course, as El Pasoans insisted (with good linguistic justification), "everyone has an accent." Their accents, which provide authenticity in relating to clients, only become a problem in promotions within the company. This selects against the heritage speakers of either English or Spanish, in favour of national standard and high register speakers of both. In upper-management positions the English-language accent is most important.

Management strategies now privilege Anglo-Americans with Spanish language skills and also Hispanics who have been promoted to high levels. In each of these groups the Spanish language carries a specific symbolic value in the internal control of the enterprise. The companies, which generally continue to be Anglo-centric at the national level, obtain local legitimation. A Mexican American operator said in Spanish, "Imaginate, un güero, vice presidente de la compañía, que gana cientos de miles de dólares al año hablando perfecto español" (Imagine, a blondie, vice-president of the company, that earns hundreds of thousands of dollars a year, speaking perfect Spanish). *Güero* is a distinctive Mexican word for light-skinned people, including U.S. whites, that goes well beyond blonde hair colour, its literal meaning. The following interview excerpts characterize quite well the rise of new managerial elites:

> One of the things we had to do was go to a convention, a meeting – they took us to a meeting complex. There was the president of the region, the vice-president of California (a Hispanic), the director. What the vice-president said was, "What I wanted was to speak about the importance that there are Hispanics," spoken in Spanish. And another vice-president, Hispanic, partly spoke in Spanish. And with fervour. There was applause, more emphatic than with other people. And he also made jokes. The presentations were partially in Spanish. This is not something normal in this country. With the Hispanic presidents, there was more charisma. (Customer service technician for a major global telecommunications company, bilingual, U.S.-born, Mexican origin)

Speaking of another major firm:

> And there was a person there who always was battling that we should speak better – my boss, an Anglo-American, but raised in Mexico City. And I thought, "How could you not listen to him? He is a brilliant man." He spoke perfect Spanish. (Training director at the call centre of a large corporation, originally from Ciudad Juárez, university educated)

It is important to keep in mind that this visibly celebrated new Hispanic or Spanish-speaking management element – still very limited – is built on a base of unrecognized, "heritage Spanish at zero cost," poorly paid, bilingual operators. Praising the value of Spanish and the importance of Hispanic talent in the U.S. market before an audience of Mexican American workers implies a reworked form of domination, in so far as this revalues "the Hispanic" led by *"güeros."* The labouring mass remains acculturated working-class Mexican Americans. Their informal-register language repertoires in both languages, but especially Spanish, are now placed below elites with formal register skills in Spanish as well as in English who are recruited among both Anglo-Americans and well-educated, often upper-class Mexicans. This apparently ethnically progressive pattern contrasts with an older pattern of domination by monolingual English-speaking elites.

We also suspect that there are subtle gender processes at work. The majority of operators were women, as were a smaller majority of language professionals and trainers; gender was thus cross-cut by educational and linguistic register differences between those two groups. In this situation of intersectionality the differences we analyse that occur among women[17] occur as often between men and women, though the presence of gendered performances cannot be ignored (as when female trainers focus male and female operators on politeness and similar topics). However, the people we interviewed about climbing career ladders were male, for reasons about which we can speculate, though this was not the focus of our research project. Corporate management, it seems to us, selected young, male, Hispanic talent as part of its market and representational strategy, compounded by gendered requirements for geographic mobility, while the geographically emplaced worksite in El Paso was relatively more feminized, though also involving many local men.

The symbolic politics of Spanish language and Hispanic ethnicity, arguably reinforced by gender, may indeed legitimize domination more effectively ("with Hispanic presidents there was more charisma"), but profound inequalities of social and linguistic origin persist, favouring carriers of nationalized formal codes in both languages over the producers of value, heritage-language speakers. Indeed, one can argue that in its invisibility and orchestration of positive affect, this is a powerful form of symbolic violence, as opposed to the visible and easier-to-resist domination by monolingual English whites in central corporate offices, which is widely recognized and resented in El Paso branch operations.

Conclusions

An important theme in the literature on call centres and language is nationalistically encoded tensions and accommodations between clients in one country and workers in other countries over language performance and ideology, often concerning varieties of English. The case of bilingual call centres based in the U.S.-Mexican border region adds new inquiries in interesting ways. In this site Latin American immigrant and second-generation clients have few linguistic performance and ideology conflicts with call centre operators (there are occasional conflicts when they are assigned monolingual English speakers from the U.S. interior). They both use informal register patterns in either or both languages, they code switch, and they use English with Spanish influences – often because they derive from the same population. This mutes nationalism and in some cases builds marketable solidarity between worker, firm, and client, what in the U.S. case has been referred to as "Latinos, Inc." (Dávila, 2001). (There are some conflicts between the subgroups of Spanish-dominant operators and monolingual English clients.) However, this positive production of value in the communicative economy goes largely unrewarded, as Spanish speech and English-Spanish bilingualism are treated as natural outcomes of the border environment, unworthy of extra compensation, training, and certification. Indeed, border speech is useful to corporations as a marker of low wages, disposability, and thus exploitability. It is an important case of the renewal of "cheap labour" in the new information economy (Alarcón & Heyman, 2013). For workers the maintenance of the Spanish language can be understood as an instrument of access to this flawed labour market, though less clearly as a means to promotion and wage increase (Cortina, Pinto, & De la Garza, 2008).

While tensions with clients are mostly subdued, call centre workers do conflict with guardians of nationalistic language purism and educated standards within their own corporations. In several important instances, border bilinguals confront nationalistic Mexican language ideologies that prioritize educated standards over their own oral, peasant-worker, and English-mixed or -influenced speech. These include everyday interactions in the border region, some specific work functions (such as text translation), and a few (but increasingly important) cases of corporate Hispanic promotion. The conflicts point towards the promising topic of world Spanishes (García, 2008; Mar-Molinero, 2004) in ways that might build on research on world Englishes, though without assuming identical patterns. Likewise, mainly English-monolingual U.S. central corporate offices distrust and stigmatize border English speakers, though that does not prevent them from hiring them. This emerges most strongly as an issue when front-line operators seek to rise into management, a fortress of ideologies of U.S. English linguistic superiority. In important instances, persons of Mexican origin in the United States are thought of as useful but low-value labour, associated with the border. In accord with García and Mason (2009), this study (also see Alarcón and Heyman, 2012) has found that immigration and the revaluation of the Spanish-language market have not resulted in changes in the unrecognized or stigmatized place of Spanish and Spanish-English bilingualism in the United States.

We can think of the situation of the border linguistic situation, and the wider border cultural situation of which it is part, in terms of "national monuments," a term derived from Renato Rosaldo's magisterial book *Culture and Truth* (1989). Rosaldo sees the border condition as occurring in all social-cultural spaces between or outside these massive, solid "monuments." However, unlike the romantic connotations of his analysis, and generally of people who read him, the border condition cannot simply be celebrated. The border bilingual call centre operators are taken for granted, poorly paid and poorly cultivated, easily disposed of, and move through a complicated field of conflict, usually in a stigmatized and powerless position, though with some performances of critique and resistance. Nor should we forget that this is not just a matter of identities and nationalistic ideologies, for also involved are capitalist relations of value transfer from workers to corporations, within which "heritage Spanish at zero cost" is crucial.

NOTES

1 *Diglossia* refers to a situation in which there are two (or more) languages or highly distinct forms of language, one being socially ranked above the other.
2 A heritage language is a language learned at home or in the community that differs from the dominant language, which is often the language used predominantly in education and institutions.
3 Register is a variety of a language used for a purpose or in a social setting; we mainly refer to the latter case, in which case registers are often ranked in terms of prestige and power.
4 *Latin@* is a neologism that includes both the female *-a* ending and the male *-o* ending by using the @ symbol (*Latin@s* is the plural).
5 There is no full account of contending linguistic ideologies on the U.S.-Mexican border, though Hidalgo (1995) and Martínez (2006) provide a start; mainly, this passage is based on more than three decades of research and residence by Heyman in the region.
6 Prosody is the patterns of stress and intonation in a language.
7 Syntax in linguistics is the arrangement of words and phrases to create well-formed utterances in a given language, approximately the same as the common word grammar.
8 Code switching is the alternation between two or more languages, often very rapidly, in a single conversation.

9 Latin American varieties of Spanish, while relatively uniform compared to the linguistic diversity of Spain, do exhibit variation between countries, regions within countries, social class, and ethno-racial group. Educated national standards are ranked above other varieties.

10 Duchêne (2009) does examine, however, cross-national interactions in several different languages in a study of a Swiss tourist call centre serving diverse countries.

11 In the company with the most prestige in the sector (State Farm), 37.5 per cent have a bachelor's degree and 34.8 per cent have an Associate's degree from a community college. In the same company, 67.5 per cent report being enrolled in university.

12 We established three call centre types for field research – five in-house, inbound; three outsourced, inbound; and one outsourced, outbound – which encompassed the main types in the area. We conducted in-depth, semi-structured interviews, distributed as described in the text. Sample selection was based on key informants' interviews and snowball procedure.

13 Additionally, in our interviews with managers other locational factors appeared. Continuing the series, other reasons include (d) the growth of education in the region, (e) high levels of available workforce (relatively elevated unemployment), (f) the growing number of call centres in place, forming a local cluster, and (g), according to the product of some companies, legal restrictions on offshoring.

14 *Cosmopolitan* refers to several possible connections. While the theme extends beyond this chapter, educated Mexicans generally and language professionals (for example, translators) in call centres specifically utilize prestige standards of two sorts: (1) Latin American and Latin North American non-national standards emerging from media corporations; and (2) the Real Academia Española, unofficial in Latin America but the official standard setter in Spain.

15 This is a telling example of drawing class distinctions among Mexicans, because *llanta* is almost universal in that country, in all registers, so that mandating the use of the highly educated *neumático* is a pointed assertion of linguistic superiority.

16 *Goma* is a Caribbean regionalism.

17 This resembles Melissa Wright's (2006, pp. 123–50) case study of Gloria, a Mexican supervisor of an export assembly plant who gained the loyalty and intensified the work of women operatives, while being excluded from the decision making of U.S. male-dominated upper management. For more on the pink-collar versus white-collar divide among women, see Poster (2002).

REFERENCES

Achugar, M., & Pessoa, S. (2009). Power and place: Language attitudes towards Spanish in a bilingual academic community in Southwest Texas. *Spanish in Context, 6*(2), 199–223. http://dx.doi.org/10.1075/sic.6.2.03ach

Alarcón, A., & Heyman, J.M. (2012). La frontera del español: Límites socioeconómicos a la extensión social de la lengua española en los Estados Unidos. *Revista Española de Investigaciones Sociologicas, 139,* 3–20.

Alarcón, A., & Heyman, J.M. (2013). Bilingual call centers at the U.S.-Mexico border: Location and linguistic markers of exploitability. *Language in Society, 42*(1), 1–21. http://dx.doi.org/10.1017/S0047404512000875

Alarcón, A., & Heyman, J.M.. (2014). From "Spanish-only" cheap labor to stratified bilingualism: Language, markets and institutions at the USA-Mexico border. *International Journal of the Sociology of Language, 227,* 101–18.

American Community Survey. (2008), http://factfinder.census.gov/faces/tableservices/jsf/pages/productview.xhtml?fpt=table. Query for "geography El Paso County, Texas, and year 2008" (accessed 7 April 2016).

Batt, R. (2002). Managing customer services: Human resource practices, quit rates, and sales growth. *Academy of Management Journal, 45*(3), 587–97. http://dx.doi.org/10.2307/3069383

Boix, E., & Vila, F.X. (1998). *Sociolingüística de la llengua catalana.* Barcelona: Ariel.

Boutet, J. (2001). Le travail devient-il intellectuel? *Travailler: Revue Internatinale du Psychopathologie et Psychodynamique du Travail, 6*(2), 55–70. http://dx.doi.org/10.3917/trav.006.0055

Boyer, H. (2006). El nacionalisme lingüístic: Una opció intervencionista davant les concepcions liberals del mercat lingüístic. *Noves SL: Revista de Sociolingüística,* Tardor-hivern, 1–12. Barcelona: Secretaria de Política Lingüística, Generalitat de Catalunya. http://www.gencat.cat/llengua/noves/noves/hm06tardor-hivern/boyer1_3.htm (accessed 1 April 2016).

Breton, A. (1998). An economic analysis of language. In A. Breton (Ed.), *Economic approaches to language and bilingualism: New Canadian perspectives* (pp. 1–36). Toronto: Department of Public Works and Government Services.

Callahan, R.M., Wilkinson, L., & Muller, C. (2010). Academic achievement and course taking among language minority youth in U.S. schools: Effects of ESL placement. *Educational Evaluation and Policy Analysis, 32*(1), 84–117. http://dx.doi.org/10.3102/0162373709359805

Cameron, D. (2000). Styling the worker: Gender and the commodification of language in the globalized service economy. *Journal of Sociolinguistics, 4*(3), 323–47. http://dx.doi.org/10.1111/1467-9481.00119

Castells, M. (1996). *The rise of the network society: The information age; Economy, society and culture* (vol. 1). Cambridge, MA: Blackwell.

Cohen, D. (2009). *Three lectures on post-industrial society*. Cambridge, MA: MIT Press.

Cortina, J., Pinto, P., & De la Garza, R.R. (2008). No entiendo: The effects of bilingualism on Hispanic earnings. ISERP Columbia University Working Paper, 2008–6.

Cowie, C. (2007). The accents of outsourcing: The meanings of "neutral" in the Indian call centre industry. *World Englishes, 26*(3), 316–30. http://dx.doi.org/10.1111/j.1467-971X.2007.00511.x

Del Bono, A. (2001). *Call centers: ¿El trabajo del futuro? Sociologia del Trabajo, 39*, 3–31.

Duchêne, A. (2009). Marketing, management and performance: Multilingualism as commodity in a tourism call centre. *Language Policy, 8*(1), 27–50. http://dx.doi.org/10.1007/s10993-008-9115-6

Duchêne, A., & Heller, M. (Eds.). (2011). *Language in late capitalism. Pride and profit.* New York: Routledge.

Fernández, L., Howard, C., & Amastae, J. (2007). Education, race/ethnicity and out-migration from a border city. *Population Research and Policy Review, 26*(1), 103–24. http://dx.doi.org/10.1007/s11113-007-9023-z

Gal, S., & Woolard, K.A. (2001). Constructing languages and publics: Authority and representation. In S. Gal & K. Woolard (Eds.), *Languages and publics: The making of authority* (pp. 1–12). Manchester, UK: St Jerome.

Ganster, P., & Lorey, D.E. (2008). *The U.S.-Mexican border into the twenty-first century*. Lanham, MD: Rowman & Littlefield.

García, O. (2008). Spanish as a global language. *International Multilingual Research Journal, 2*(1-2).

García, O., & Mason, L. (2009). Where in the world is U.S. Spanish? Creating a space of opportunity for U.S. Latinos. In W. Harbert, S. McConnell-Ginet, A. Miller, & J. Whitman (Eds.), *Language and poverty* (pp. 78–101). Clevedon, UK: Multilingual Matters.

Gibson, E.K., Olmedo, C., & Caire, M.E. (2008). *City of El Paso: Cross sectional comparison of bilingualism in the workplace*. IPED Technical Report No. 2008-07. Institute for Political and Economic Development, University of Texas at El Paso.

Grenier, G. (1982). *Language as human capital: Theoretical framework and application to Spanish-speaking Americans*. Doctoral dissertation, Princeton University.

Grin, F. (1999). The notions of supply and demand as analytical tools in language policy. In A. Breton (Ed.), *Exploring the Economics of Language* (pp. 31–61). Ottawa: Official Languages Support Program, Canadian Heritage.

Grin, F., Sfreddo, C., & Vaillancourt, F. (2010). *The economics of the multilingual workplace*. New York: Routledge.

Heller, M. (2003). Globalization, the new economy and the commodification of language and identity. *Journal of Sociolinguistics, 7*(4), 473–92. http://dx.doi.org/10.1111/j.1467-9841.2003.00238.x

Heller, M. (2010). The commodification of language. *Annual Review of Anthropology, 39*(1), 101–14. http://dx.doi.org/10.1146/annurev.anthro.012809.104951

Heskett, J.L., Sasser, W.E., Jr, & Schlesinger, L.A. (1997). *The service profit chain: How leading companies link profit and growth to loyalty, satisfaction, and value*. New York: Free Press.

Heyman, J.M. (2012). Constructing a "perfect" wall: Race, class, and citizenship in US-Mexico border policing. In P.G. Barber & W. Lem (Eds.), *Migration in the 21st century: Political economy and ethnography* (pp. 153–74). New York: Routledge.

Hidalgo, M. (1995). Language and ethnicity in the "taboo" region: The US Mexican border. *International Journal of the Sociology of Language, 114*(1), 29–46. http://dx.doi.org/10.1515/ijsl.1995.114.29

Hudson, A., Cháves, E.H., & Bills, G.D. (1995). The many faces of language maintenance: Spanish language claiming in five southwestern states. In C. Silva-Corvalán (Ed.), *Spanish in four continents: Studies in language contact and bilingualism* (pp. 165–83). Washington, DC: Georgetown University Press.

Kelly-Holmes, H., & Mautner, G. (2010). *Language and the market*. Basingstoke, UK: Palgrave Macmillan.

King, K.A. (2009). Global connections: Language policies and international call centers. *Language Policy, 8*(1), 1–3. http://dx.doi.org/10.1007/s10993-008-9123-6

Kolenc, V. (2010). Recession-proof: El Paso's call-center industry thriving and hiring. *El Paso Times*. Retrieved from http://www.elpasotimes.com/news/ci_16171721.

Lago, E. (2009). *Estados Unidos Hispanos: Enciclopedia del español en los Estados Unidos*, Coordinated by Humberto López Morales. Madrid: Instituto Cervantes-Santillana.

Leidner, R. (1993). *Fast food, fast talk: Service work and the routinizations of everyday life*. Berkeley: University of California Press.

Mar-Molinero, C. (2004). Spanish as a world language: Language and identity in a global era. *Spanish in Context, 1*(1), 3–20. http://dx.doi.org/10.1075/sic.1.1.03mar

Martínez, G.A. (2006). *Mexican Americans and language: Del dicho al hecho*. Tucson: University of Arizona Press.

Milroy, J., & Milroy, L. (1998). *Authority in language: Investigating Standard English*. London: Routledge.

Mirchandani, K. (2004). Practices of global capital: Gaps, cracks and ironies in transnational call centres in India. *Global Networks, 4*(4), 355–73. http://dx.doi.org/10.1111/j.1471-0374.2004.00098.x

Mirchandani, K. (2005). Gender eclipsed? Racial hierarchies in transnational call center work. *Social Justice (San Francisco, Calif.), 32*(4), 105–19.

Montejano, D. (1987). *Anglos and Mexicans in the making of Texas, 1836–1986.* Austin: University of Texas Press.

Morgan, B., & Ramanathan, V. (2009). Outsourcing, globalizing economics, and shifting language policies: Issues in managing Indian call centres. *Language Policy, 8*(1), 69–80. http://dx.doi.org/10.1007/s10993-008-9111-x

Moss, P., Salzman, H., & Tilly, C. (2008). Under construction: The continuing evolution of job structures in call centers. *Industrial Relations, 47*(2), 173–208. http://dx.doi.org/10.1111/j.1468-232X.2008.00516.x

Poster, W.R. (2002). Racialism, sexuality, and masculinity: Gendering "global ethnography" of the workplace. *Social Politics, 9*(1), 126–58. http://dx.doi.org/10.1093/sp/9.1.126

Poster, W.R. (2007). Who's on the line? Indian call center agents pose as Americans for US-outsourced firms. *Industrial Relations, 46*(2), 271–304. http://dx.doi.org/10.1111/j.1468-232X.2007.00468.x

Rahman, T. (2009). Language ideology, identity and the commodification of language in the call centers of Pakistan. *Language in Society, 38*(02), 233–58. http://dx.doi.org/10.1017/S0047404509090344

Read, B.B. (2002, 7 January). Star-spangled site selection. *Call Center Magazine.* Unpaginated, files of the authors.

REDCO [El Paso Regional Economic Development Corporation]. (2010). Call center statistics. (Unpublished manuscript). El Paso: REDCO.

Rojas, Y. (2007, March). Call centers in Northeast Mexico. Washington, DC: U.S. Department of Commerce.

Rosaldo, R. (1989). *Culture and truth: The remaking of social analysis.* Boston: Beacon Press.

Roy, S. (2003). Bilingualism and standardization in a Canadian call center: Challenges for a linguistic minority community. In R. Bayley & S. Schecter (Eds.), *Language socialization in bilingual and multilingual societies* (pp. 269–85). Bristol, UK: Multilingual Matters.

Schieffelin, B., & Woolard, K.A. (1994). Language ideology. *Annual Review of Anthropology, 23*(1), 55–82. http://dx.doi.org/10.1146/annurev. an.23.100194.000415

Silverstein, M. (1979). Language structure and linguistic ideology. In P. Cylne, W. Hanks, & C. Hofbauer (Eds.), *Elements; A parasession on linguistic units and levels* (pp. 193–247). Chicago: Chicago Linguistic Society.

Sonntag, S.K. (2009). Linguistic globalization and the call center industry: Imperialism, hegemony or cosmopolitanism? *Language Policy, 8*(1), 5–25. http://dx.doi.org/10.1007/s10993-008-9112-9

Spener, D. (2002). The unraveling seam: NAFTA and the decline of the apparel industry in El Paso, Texas. In G. Gereffi, D. Spencer, & J. Bair (Eds.), *Free trade and uneven development: The North American apparel industry after NAFTA* (pp. 139–60). Philadelphia, PA: Temple University Press.

Spolsky, B. (2009). *Language management*. Cambridge, UK: Cambridge University Press. http://dx.doi.org/10.1017/CBO9780511626470

Tan, P.K.W, & Rubdy, R. (2008). *Language as commodity: Global structures, local marketplaces*. London: Continuum.

Taylor, P., & Bain, P. (2005). "India calling to the far away towns": The call centre labour process and globalization. *Work, Employment and Society, 19*(2), 261–82. http://dx.doi.org/10.1177/0950017005053170

Timmons, W.H. (1990). *El Paso: A borderlands history*. El Paso: Texas Western Press.

Urciuoli, B. (2008). Skills and selves in the new workplace. *American Ethnologist, 35*(2), 211–28. http://dx.doi.org/10.1111/j.1548-1425.2008.00031.x

Urciuoli, B., & LaDousa, C. (2013). Language management/labour. *Annual Review of Anthropology, 42*(1), 175–90. http://dx.doi.org/10.1146/annurev-anthro-092412-155524

Velázquez, I. (2009). Intergenerational Spanish transmission in El Paso, Texas: Parental perceptions of cost/benefit. *Spanish in Context, 6*(1), 69–84. http://dx.doi.org/10.1075/sic.6.1.05vel

Veltman, C. (1988). *The future of the Spanish language in the United States*. New York: Hispanic Policy Development Project.

Vila, P. (2000). *Crossing borders, reinforcing borders: Social categories, metaphors, and narrative identities on the U.S.-Mexico frontier*. Austin: University of Texas Press.

Villa, D., & Villa, J. (2005). Language instrumentality in a border region: Implications for the loss of Spanish in the Southwest. *Southwest Journal of Linguistics, 24*, 169–84.

Wiley, T.G. (2001). On defining heritage learners and their speakers. In J. Peyton, D. Ranard, & S. McGinnis (Eds.), *Heritage languages in America: Preserving a national resource* (pp. 29–36). McHenry, IL: Delta Systems.

Wirtz, R.A. (2001, May). Phone home. *fedgazette*. Minneapolis Federal Reserve Bank. https://www.minneapolisfed.org/publications/fedgazette/phone-home (accessed 9 April 2016).

Wright, M. (2006). *Disposable women and other myths of global capitalism*. New York and London: Routledge.

Summary

9 Nations at Work in Transnational Call Centres

KIRAN MIRCHANDANI AND WINIFRED R. POSTER

With the chapters in this volume we have sought to initiate a dialogue between those studying labour and those exploring nations. Scholarship on the nation has largely overlooked the ways in which nationalism is exercised as part of many people's day-to-day jobs. Even in analyses of labour in war industries such as aircraft and weapon manufacturing, which are so closely tied to the maintenance of national boundaries, the focus is on the construction and sale of commodities rather than on the ideological production of nations. Professions in, for example, the army and immigration control are seen to be related to the enforcement of laws rather than to the performance of labour that results in the creation of nations. Nation-related activity is assumed to be limited to public sector jobs that focus on political leadership or the creation and implementation of policies around immigration and foreign affairs.

In the same way, while there is a growing recognition of the centrality of workers' emotional labour within studies of work, there has been little focus on how this labour not only facilitates relationships with customers and the formation of identities but also plays an important role in the creation of nations. The contributions in this collection attest to the labour involved in everyday nationalism, explored through the empirical case of the transnational call centre sector. No doubt, nations and their borders are also being lived and given meaning by a variety of other types of globalized workforces, such as transnational domestic and care workers who play a significant role in the socialization of children in wealthy nations; global executives who work closely with one another in cross-national teams; manufacturing workers who are employed in foreign-owned enterprises; and state officials whose domains are touched by foreign policy. Indeed, in today's global economy, doing one's job often involves constructing one's and others' nations. Voice-to-voice transnational interactions in call centres makes these particularly interesting sites where the labour of work and the labour of nation are intertwined, as the ethnographies in this collection have illustrated.

In this concluding chapter we trace some of the central ways in which the work of the construction of nations and the work of doing one's job overlap in the call centre sector globally. Ethnographies from call centres around the world help to debunk popular and scholarly assumptions about how global outsourcing operates. One such commonly held assumption is that technological standardization leads to transportability of the labour process, and, in turn, unfettered geographical mobility, so that transnational firms can be set up anywhere. Yet, even though call centres in different countries may use the same telephone systems, and calls may be transmitted through nationally generic telephone lines, local industries differ dramatically from one another. Worker positions, aspirations, and orientations determine the ways in which employees perform the emotional labour of service. This confirms that service encounters involve the exchange of information in embodied ways. In the process the transnational call centre sector may facilitate or frustrate aspirations of individual workers. Beyond that, however, it is also fundamentally intertwined with national aspirations as constructed by global citizen workers, organizational decision makers, and state bureaucrats.

In particular, the chapters in this collection point to the central role of states and other call centre actors in conceptualizing nations in relation to the international. While transnational organizations may seem placeless, they in fact comprise actors who are deeply embedded in and committed to national politics. Workers' daily lives depend on nation-based binaries between customers and servers. In fact, the very justification for geographical proliferation of the industry is the existence of labour-cost hierarchies across different nations. Far from dissolving boundaries, the transnational call centre sector involves the continual expression, particularly by state actors, of hierarchies between nations. International exchange among those engaged with outsourced call centres remains firmly entrenched within the discourse of national competition, political conflict, and territorial dispute. As such, global citizen workers in other countries are imagined as threats to national economic sovereignty. Yet, as this collection attests, workers express similar challenges across contexts. This commonality suggests a ripe (although thus far largely untapped) opportunity for cross-nation worker advocacy.

De-centring the Global Service Economy

At the outset a prominent theme of this volume is the considerable shift in the geography of the global call centre industry over the last three decades. The focus of much of the early scholarship on this sector was a handful of developed countries (the United Kingdom, the United States, Australia, and Canada). In the early 2000s, India emerged as a new site for both the location and the study of outsourced customer service work. Evidence in this volume suggests, however, that the outsourcing of customer service work is now a more multifaceted and multi-sited phenomenon. We trace seven countries across five continents to show how global call centres pervade varying national landscapes, especially in the non-Western and non-northern world. Indeed, we expect that the information in this book represents the tip of the iceberg rather than a comprehensive sweep of the phenomenon.

One of the remarkable points here is the *range of economic contexts* in which global call centres operate. Many of these defy expectations about typical global mappings. For instance, in some cases, like that of Mauritius (chapter 4), call centres are situated in relatively well-off and even egalitarian settings of the industrializing world. Its labour force, moreover, is almost entirely represented by collective bargaining. Similarly, transnational call centres in some nations are promoted (though not always experienced) as providing skilled, professional, clean, white-collar jobs. The sector has not only economic but, more important, symbolic power in countries like Mauritius and Morocco. In such places, lavish organizational spaces serve to produce an upscaled image of the sector within the nation. Rather than being constructed as just "cheap," ideal workers are English-speaking cosmopolitan youth who are encouraged to aspire to middle-class lifestyles.

Heyman and Alarcón (in chapter 8) reveal the opposite case in their account of the U.S.-Mexican border – the hidden pockets of labour deprivation within what has been called the wealthiest country of the world. Call centre workers may earn higher wages in the United States compared to many other countries in the context of minimum wage legislation, but those located in El Paso, on the border of Mexico, earn among the lowest in the United States, with hourly earnings of two to four dollars below the national average for the same occupation. This is a region with a high population of Latinos. As such, the authors' findings confirm a pattern well-documented in other studies of labour among U.S. information and communication technology industries – a search for

the lowest common denominator in labour standards. High-tech industries engage in these geographic dynamics of exploitation as well; they
often move to the cheapest possible workforce, which they conceive of
as immigrant women and women of colour. Examples include computer
electronics assembly in Silicon Valley (Hossfeld, 1990) and semi-conduc-
tor factories on Native American reservations (Nakamura, 2014).

Sociologist Robert Blauner (1972) referred to this process as "internal colonialism," as global north countries recreate conditions of global
south labour, within specific regional mappings, in their own borders.
Scholars of rural call centres in the United States have noted this pattern
as well. In an industry highly concerned with expenses of labour and
rent, the call centre sector has expanded in rural areas (like the north-
west) to reduce costs. It has achieved this by hiring largely rural, female
workforces in part-time jobs at wage rates that are much lower than
those in urban areas (Bonds, 2006). That this regional exploitation of call
centre workers is happening on the Mexican border is a testament to the
devaluation of Latino workers and Spanish-language skills – despite the
fact that those skills are in increasing demand.

A second sign of decentring in the global call centre industry is the
shifting locus of *where firms are going*. India spearheaded the outsourcing industry but is now losing its monopoly as an offshore destination.
Some analysts claim that countries like the Philippines currently host
equal if not larger numbers of call centre workers (Bajaj, 2011). Furthermore, many other territories now compete for Indian contracts, such as
those in our volume: Guyana, Mauritius, and the El Paso region of the
United States.

However, to view this trend as a diminishment of India's role in the
overall call centre industry would be misleading. Indeed, chapters in
this volume cue us to another important story – how India has adapted
to become a key facilitator of the industry in other global south countries. We see how India has transformed from being an outsourcing
provider to becoming a multinational contractor of services in other
countries. In particular, entrepreneurs are establishing call centre firms
in countries such as Guatemala (chapter 6) and the Philippines (chapter 5) to receive and fulfil service customer contracts from the United
States.

More important, recent studies show that India sometimes has considerable authority in these contexts (Mann, Graham, & Friederici,
2014). Indian entrepreneurs play critical roles in a variety of global centre functions outside its borders, namely in (1) setting up industries

in other countries, through expertise and consulting; (2) gatekeeping client contracts, through their "re-outsourcing" of services from the global north, which they can choose to transfer to places like East Africa, or not; and (3) downgrading the content of projects in the process, by passing on lower-level call centre work to countries in the global south and keeping higher-level functions for itself.

In this capacity India now has key resources in its diasporic ties to the United States and the United Kingdom for obtaining contracts (which can then be transnationally dispersed) and in its knowledge of how to nurture industries. Furthermore, it has become a hub both for training (as managers and workers from around the world travel to sites in India for this purpose) and for technology (as the digital campaigns themselves go from countries in East Africa through Indian technological infrastructures, like VOIP). As Mann, Graham, and Friederici (2014) reveal, having ties to India can be the make or break determinant of survivability for some call centres in Kenya and Rwanda. This represents the emergence of Indian entrepreneurs on the global stage, through a techno-masculinity that subverts traditional north-south hierarchies, as they become primary agents of information and communication industries outside their own national borders (Poster, 2013b).

Towards this end, a third marker of decentring in the call centre industry is a re-siting of its *broader organizational infrastructure*. A salient trend is the movement of call centre administration and resource centres to the global south. A plethora of non-Western and non-northern countries are now involved in the wider circle of agents that are developing the industry. In East African call centres, for example, resources are being pulled from places as diverse as China, Cambodia, Egypt, and Dubai (not to mention Japan) for things ranging from service contracts to venture capital to hardware (Mann et al., 2014).

All of these trends, therefore, including the relocation of outsourcing destinations, reflect what theorists are calling the rise of "south-south" dynamics in the global system. The call centre industry is appearing to reconsider its exclusive dependence upon the global north. While the majority of the service contracts still emerge from countries in North America and Europe, we are also seeing a rise in regional systems of outsourcing within the global south itself (see Qiu, 2010, on the Chinese call centre industry, for instance).

Finally, an examination of remappings in the call centre industry reveals the *endurance of transnational hierarchies*. Theories of a borderless world would predict that all countries should have similar, if not equal, opportunities to participate in the new-found information economy. Technological infrastructures and networked communications should enable any nation to become a site for customer service outsourcing. However, recent case studies are showing that this is not the case. Where outsourcing *doesn't* go, rather, is highly indicative of dynamics in the neoliberal economy.

Rwanda and Kenya are cases in point (Mann et al., 2014). They recently acquired access to fibre-optic cables and affordable Internet broadband, starting in 2009. They also have English-speaking, educated workforces and, by some accounts, wage rates that are lower than those of India. Yet neither cost savings nor English-language capabilities appear to be sufficient factors to draw foreign clients to their firms. Local entrepreneurs attempting to set up international call centres recount a range of nation-based stereotyping (that is, in management language, a "lack of trust") by foreign client firms.

This reflects the presence of clearly demarcated preferences for global south destinations by firms in the global north, in which East African nations, for instance, are ranked among the lowest. It confirms what our post-colonial framework alluded to in the introductory chapter: the continuing relevance of nation-specific imageries of outsourcing destinations, set within a stratified global system. In direct contradiction to the theorizing of Friedman's view of global equalizing, as Mann, Graham and Friederici (2014) articulate so aptly, it reveals a "non-flat world."

In sum, these global patterns – of geographic proliferation, expansion, and selectivity – encourage a rethinking of the knowledge gathered to date on the call centre sector. They suggest that, despite the near universal adoption of computer-based dialing technologies, performance mechanisms, and labour-process protocols in the industry, there is in fact considerable diversity in the ways in which call centre work is experienced.

Patterns and Sources of Banal Nationalism in Call Centres

As stated in the introduction, our prism for exploring nationhood and national identity in the global call centre industry is the workings of banal nationalism (Fox & Miller-Idriss, 2008). Chapters here have documented the way that "nation" is created on a daily basis through everyday acts and routines, from formalized organizational policies to the casual conversations of workers. Of particular importance is the way that statehood is unhinged from its geographic moorings and in fact created as an "imagined community" (Anderson, 1991). In the case of Mauritius (chapter 4), call centres literally choose from a variety of national affiliations to create an acceptable identity to appeal to their consumers and their corporate clients. They take advantage of their "multiplex" status and proximity to many states to become whatever nation suits a particular caller: workers may identify their locations as "off the coast of India" to customers from some countries, or "off the coast of South Africa" to others, or even "off the coast of Australia" to yet others. Critical in this regard is the multi-staged process of conceptualizing and verbalizing nationhood. Workers are creating place by *dis-identifying* with undesirable countries (for example, India, for Americans) and then *re-identifying* with others (for example, Australia). National affiliation becomes changeable, and formulated situationally, through transnational interactions.

Our intention at a basic level is to show that call centres cannot be fully understood without attention to their contexts – local, national, and global. However, we also want to convey that the banal nationalism of call centres operates on two fronts, or in two directions. On the one hand, states (and their associates) assert their political and economic agendas through the call centre industry, especially in the way they promote or denigrate its workforce. On the other hand, call centre officials (including their administrators, employers, and managers) benefit from deploying a sense of nationhood and nationality for their workers. The deeply nuanced ethnographies of this volume, therefore, document how call centres (and their workers) mediate between nations in the global economy and, in fact, are part of the process used to define place.

The banal nationalism (Billig, 1995) of call centres has many sources. Chapters in this volume document a range of factors in the regional, national, and global contexts that serve as aids in, and sometimes pressures for, the process of imagining the nation. Some of them, intuitively, involve the *state*. As many of our chapters show, state governments are engaging in direct and indirect support of the call centre industry, and shaping its workforce towards its aims. This is the case in the Philippines, Guatemala, Guyana, and El Salvador (chapters 5, 6, 3, and 2, respectively), where state offices actively create policies for the recruiting of employees to call centres, the advertising of call centres to corporations overseas for contracts, and much more. In Mexico government officials are reported to hand out information about call centre job opportunities to deportees as they cross the border (Wessler, 2014). Call centre firms, in turn, follow the settlement patterns of deportees and relocate to neighbourhoods in which those potential workers are living.

States, however, do not typically act alone in moulding the call centre industry. Quite often they are networked with a variety of ancillary agents and stakeholders, locally and abroad, who have their own purposes and interests in its success. For instance, in chapters 2 and 3, we see that the *tourist industry* is closely aligned with call centres in the eyes of the state. Tourism has been long intertwined with the economic histories of countries like Guyana and El Salvador as a means for courting private capital from abroad. Furthermore, tourism has fed into and helped fuel the call centre industry by preparing workers with proper accents and cosmopolitan social capital. Some tourism-related service firms are even located in special economic zones with global call centres. Consequently, these organizations have a stake in the success of call centres because they benefit from the circulation of foreign personnel arriving in these locations.

Another formative group for the call centre industry is the cadre of *overseas entrepreneurs*. Benner and Rossi (chapter 4) illustrate the key role of diasporic business networks in facilitating transnational corporate investments. In Mauritius, for example, these networks span India, China, France, the Middle East, and the United Kingdom, and many transnational business opportunities have been facilitated through personal, family, and social networks. These linking individuals become the "strategic agents" (Sassen, 1996) who both serve global capital and represent the nation's aspirations.

Recent to the scene as well, and growing in their influence, are the *transnational banking and lending agencies*. Organizations like the World Bank, international aid groups, and grant foundations are now looking to call centres as a development strategy for global south countries (Mann et al., 2014). No longer just an exercise for entrepreneurs, call centres are part of the neoliberal regime of global-poverty alleviation and debt financing. In accordance, seventy countries have listed call centres in their "Vision 2030" plan (Graham & Mann, 2013). Not only are finance organizations urging global south states and entrepreneurs towards this industry, but they have material interests in the outcomes.

Lastly, but very significantly, a source of banal nationalism is the *workers* themselves and their capacity to draw from, produce, and exhibit certain kinds of global social capital. Some of these workers are local residents but with global outlooks and orientations.

For instance in Guyana and the Philippines (chapters 3 and 5), call centre administrators have been turning to local youth who are aspiring to move into middle-class positions. These workers are often enticed by a promise of high-paid, white-collar work, a promise that may prove unfulfilled for many. Other kinds of workers are transnational migrants. By hiring workers from outside their borders, call centre employers acquire "nationally appropriate" workforces by default and without having to commit resources for training. Two kinds of transnational migrants appear in this volume. One is the group of foreign workers who come from the country of the corporate sponsor and its customer base. This is the case for Morocco (chapter 7), where call centres hire immigrant employees from Spain and other European countries. Elmoudden refers to this as "inverse migration," noting a counter-trend to the typical south-to-north migration – one of north to south. The second group is that of "returnees" – employees who have left their native lands, lived in the country in which the customers and corporate clients are located, and then returned. This trend is evident in Morocco (chapter 7) as well as Guatemala (chapter 6) and El Salvador (chapter 2).

In these ways call centres are capitalizing on the unique resources of mobile citizenries. Certainly, many kinds of global industries rely upon immigrant labour forces (for example, multinational firms for cheap and desperate workers, and households for immigrant care workers). What we see here, however, is a peculiar dynamic in which employers in the global south seek specific kinds of social capital from their employees that is attained by migration and global orientation – through their residence and embeddedness in countries of the global north, very often their prior colonial masters. This experience provides critical communicative skills for interacting with customers in those locations. Foreigners and cosmopolitan locals, therefore, are especially valued for their knowledge of the customer side of the interaction. Diasporic returnees (whether coming back voluntarily or through deportation) are valued for dual skills: their familiarity with the customer experience and their capability to perform national identity requirements for the homeland.

As described in the introduction, such mobile workforces represent "hybrid" employees. They offer the firm a dualized type of social capital according to nation. Indeed, recruiting this workforce aids the labour process of global call centres, sometimes more than managers may know. In a telling example, Wessler (2014) describes the case of

Angel, a returnee in Mexico who grew up in small-town Texas along "with redneck boys." Indicative of blending geographies, Angel has adopted elements of this stereotypically white, rural, and conservative identity in the United States. On the telephone, he bonds instantly with many like-minded callers, like Jim from Alabama who calls him "Bubba" and with whom he shares experiences about hunting, fishing, and river swimming. Given the audio context of the communication technology, Jim probably never knew Angel was Chicano; yet their connection most likely enhanced Jim's experience of the service and, in turn, the end goals of the call centre. Such particularized forms of American and Mexican identity emerge from the unpredictability of individual life histories in a transnational world – ones that would not likely be taught in a call centre training class.

Overall, then, this volume reveals a variety of agents and beneficiaries of banal nationalism in global call centres. Some of these are directly visible in the day-to-day operations of call centres. Others are more behind the scenes in their influence but are nonetheless significant in their path and development. Remarkably, while the research on call centres (even in comparative studies) has predominantly focused on the internal workings of the labour process, and the role of on-site managers, entrepreneurs, and administrators therein, we highlight the wide array of external forces and actors that have direct bearing on this process. These diverse sets of participants show that the intersection of capital with nation is a very textured, layered, and patchworked affair.

Themes of Labour and Nation

We trace four themes that capture the overlap between labour and nations and run through the case studies in this book: (1) the spread of national identity management, (2) the salience of nation branding, (3) the creation of the global citizen worker, and (4) the implications of transnational communication hierarchies. We derive some of our theoretical groundings from the vast literature on India, as it is the most studied player for outsourced call centres in terms of issues of labour, history, and nationalism. Yet our purpose is to show how other countries demonstrate the varied forms of constructing nations through call centre work. Indeed, the ethnographic case studies here represent a small fraction of the settings within which transnational call centres have been established in the last decade.

The Transnational Reach of National Identity Management

As mentioned in the introduction, the earliest research on nation-hood in call centres emphasized the practices of constructing nation-ally appropriate workforces, that is, employees who would mirror the characteristics of the consumer's nation. It began in the nascent years of the industry as "national identity management" (Poster, 2007). Later, parts of the regime adapted to a subtler process of "authenticity work" (Mirchandani, 2012). Rather than asking workers to state a false identity directly, managers have moved to a subtler and more covert practice of enacting Americanness through recruitment and training, demeanour and scripting, and office ambiance and architectural decor. It involves seemingly contradictory tasks: for workers, being Western and Indian at the same time; and for managers, administering outsourcing for U.S. firms but also hiding it from U.S. consumers and responding to their backlash.

While we were embedded in this fieldwork, many of us were wonder-ing whether such explicit acts of nation-making were specific to India. With this volume, we answer that question with a resounding *no*. A sig-nificant finding of this volume is the similarity of the practices to those of other countries. Take the example of the foreign alias. In Guatemala workers take on American names like Donald (chapter 6). In Morocco workers use French aliases like Marianne (chapter 7). Additional guide-lines of national identity management are found in Morocco as well. For instance, the environment inside the call centre is wiped clean of local influences, and workers are not allowed to speak Arabic on the shop floor. Workers use a script to mask nationhood when customers question their identity and location, and state that their "nationality is French of Arab descent." Such evidence indicates that the drive to shape workers in a national image is not an isolated event in India, but rather a commonplace and integral feature of outsourced service labour worldwide. Nation is something that many global call centres do.

This collection also documents the varied contours of locational masking in a global industry that has moved beyond India. Filipino workers (chapter 5), for instance, are required to be quite open about their geographical location and to promote the unique national identity of Filipinos as naturally disposed to call centre work because of their presumably calming voice intonations. Guyanese workers (chapter 2), similarly, are encouraged to let callers know that they are not in fara-way India but close to customers and within the Americas. Contrary to earlier patterns, then, a deliberate unmasking of location is sometimes considered to be a valuable asset for call centres. Furthermore, this happens directly in reference to India's role as a more established outsourcing centre. In the second decade of this industry, managers are not necessarily hiding the site of the call centre but are pointing out its contrasting (and improved) location vis-à-vis that of competitors.

Moreover, another significant finding is that workers often enact many different nationalities at once. Some call centres accept contracts from multiple nations at the same time and, in turn, ask workers to adjust their personas to those different national customer bases. Poster (2007) saw a hint of this in India, where workers were changing their names and accents when talking to customers from the United States, Australia, and the United Kingdom (the latter alone requiring sixteen different accents). Furthermore, by virtue of automatic routing systems on their phones, workers were asked to switch their identities – quickly and on demand – for successive calls from different countries.

Even across the diversities of countries represented, however, this process involves one common language. What we are starting to see now is the growth of multilingual sites. In Central and South America, countries handle both Spanish and English. In Mauritius and Morocco, call centres handle contracts from English, French, German, and Spanish countries. This trend is on the rise, as call centres regularly become globally compressed sites.

Employees continue to have many kinds of reactions to national-identity posing, as previous call centre research shows. Some workers enact Americanness voluntarily, like those in Mexico, as a way to link with and relive their former migratory homes (Wessler, 2014). In contrast, Guyanese employees (chapter 3) express objection to the affronts to their own national identity, and especially to the xenophobic abuse from customers. These dynamics are reflective of the multiplicative and hybrid identities discussed in the introduction.

Nation Branding

Branding and advertising strategies, which are traditionally used to market products, have come to play a key role in marketing nations on the global stage. As Fan (2010) notes, a brand is "a complex bundle of images, means, associations and experiences in the minds of people," and a nation brand by extension is the "total sum of perceptions of a nation in the minds of international stakeholders" (p. 98). In the context of economic

globalization the targets of this marketing are international stakeholders, including foreign tourists and global investors. Like corporate and
product brands, nation brands are constructed by professional branding strategists and circulated via the media. These strategists often
argue that the mass consumption of brands depends on the simplicity
of the message, as well as the promotion of stereotypes about nations.
Scholars observe that "when a country has a clear, simple, well-defined
national stereotype, the media will be more comfortable covering that
country" (Anholt, 2009, p. 178).

India has been at the forefront of nation-branding strategies since
the early 1990s. Mehta-Karia (2011) characterizes nation branding as
a "tool of neocolonial governmentality" (p. 7). She traces the planning
undertaken by the India Brand Equity Foundation, a public-private
partnership, for India's four-million-dollar "Brand India" strategy in
the mid-1990s, which was launched with great extravagance and fan-
fare at the World Economic Forum in 2006. The occasion was used to
signal India's "break" from its past and its emergence as a "new" nation.
New India was branded as a country that had embraced neoliberal policies and was poised to become a global economic powerhouse, teeming
with potential customers and workers.

Chapters in this volume elevate our understanding of this branding
process in other national settings. Branding operates on many targets:
nations, firms, and workers. Rivas (chapter 2) describes the advertising
of subjects within the nation through the "El Salvador Works" slogan,
which constructs local human capital as both hardworking and bilingual.
Branding Mauritius involves an overt restating of the nation's territorial
geography as "off the coast of Australia" despite its territorial proximity to Africa (Benner and Rossi, chapter 4). In this way, call centres take
advantage of the higher labour prices that companies are willing to pay
for nearshoring over offshoring. Branding Kenya involves comparisons
to other markets around the world, with its preferable location given the
"saturation of markets in India and the Philippines" (Mann et al., 2014).

Sometimes call centre brands of various countries look similar. Indeed,
the overlaps between nation-branding projects are far from accidental
because brand-making is a service provided by emerging global consultants. Yet, Fan (2010) distinguishes between national branding and
national identities and points out that, while the former is a marketing strategy constructed by professionals, the latter is under continual
definition by local residents within a territory. Indeed, as Poster (2007)
shows, an integral part of work in Indian call centres involves national

identity management, defined as organizational strategies through which "ethnicity and citizenship are considered malleable and subject to managerial control" (p. 273). Here we note the interactive relationship between states and organizations in this practice. Indeed, with the growing collaboration between state neoliberalism and global economic capital, there are parallel trends in the governmental branding of citizens, on one hand, and the organizational manipulation of these citizen brands for shareholder profit, on the other. As Trotz, Mirchandani, and Khan show (chapter 3), Guyanese call centre workers are constructed as both global middle-class citizens at the forefront of the country's economic future, and cheap labour for multinational corporations.

Substantively then, nation branding may not always have universally consistent patterns. Despite marketing attempts to convey a clear branding message, multiple and contradictory branding discourses often exist across call centre sectors in different nations. For example, while call centre employees in El Salvador, Morocco, Mauritius, and Guatemala are seen to be doing professional, middle-class, global work (much like workers in India, see Noronha and D'Cruz, 2009), Filipinos are branded as national heroes precisely because they patiently and successfully engage in demeaning international work.

Branding also situates holders in a customer-provider relation with in-built power disparities by nation. Products manufactured in nations that have positive brand associations, for example, are often valued more highly. A case in point is Germany, which is said to have high brand "equity"; its products are often more expensive and assumed to be of better quality (Kotler & Gertner, 2002). Kotler and Gertner illustrate the complex global-local networks that serve to generate such national brands, by tracing decision-making processes within a large transnational company opening a branch in Latin America. This firm selected Costa Rica as a site because of its strong brand. The national brand was constructed by consultants at the state-supported Costa Rica Investment Promotion Agency, who worked in conjunction with similar agencies in Europe, as well as local business professionals and the dean of the business school in Costa Rica. Branding, in this way, was a critical part of the marketing strategy to potential global investors, who themselves shop like customers for an appropriate location in which to make their investment. As Mehta-Karia (2011) notes, nation branding elevates the West and represents the "normalization of the overriding logic of the market ... so that the nation can now only be legitimately imagined through the language of the market" (p. 45).

Nation brands may be primarily directed at international investors and tourists, but they have domestic impact because local populations are required to "live" the brand in order to be named appropriate citizens (Mehta-Karia, 2011, p. 44). The term has a dual meaning – to brand is both to promote and to stigmatize. Hoenig (2012) describes the violent and widespread practice of branding during slavery in the 1800s, which was used to identify, to punish, and to assert ownership over individuals deemed to be slaves. While branding in the business investment literature is constructed as positive marketing, it continues to involve a similar process of naming and labeling. Implicit in the promotion of a nation brand is a devaluation of individuals and activities that fall outside its purview, a third theme that runs through this collection.

Globalized Citizen Workers

As noted in the introduction, the nationalized character of labour has often been viewed through the concept of the citizen worker. Here, we update this concept for the transnational service industry of the twenty-first century.

Our authors describe a *global* citizen worker who has a new set of qualifications. Call centre employees are now expected to (1) interact on a transnational stage and (2) present a cosmopolitan savvy to foreign consumers and potential corporate clients. At the same time, of course, they are still "doing nation" in the various ways described earlier. Hence, this apparent contradiction is reflected in another qualification of the global citizen worker: (3) to inhabit an appropriate national image while presenting this cosmopolitan display. Indeed, chapters in this volume reveal the way that call centres are constructing not only nationalized workforces but those that have particular abilities to perform and enact markers of a global citizenry.

Within our collection there are many permutations and examples of the ways that this happens. Sometimes the premise of the global citizen worker is to be transnationally *neutral* (however that is conceptualized). At times this cosmopolitanism is achieved easily for call centre employers, while at other times it requires more effort on the part of the firm. Another premise is to be *heterogeneous* in national identities (for example, posing as one nationality in the first call, then another in the next). Or, as a third example, global and national roles may be directed at different targets: one is outward for the world, while the other is inward for the locality. Here, global citizens foster an appeal to a pan-national clientele (including consumers, shareholders, and tourists), yet enacting discrete nationalities for specific audiences. They may have a cosmopolitan veneer, while being nationally branded in practice.

In all, there is a common baseline: to multi-task the national and the transnational. These two tasks may seem oppositional, but they can be quite synchronous. Significantly, one does not come at the expense of, nor compromise, the other. As Meoño Artiga describes in chapter 6, "workers do not compete to be *less* Guatemalan but to be *new* Guatemalans: flexible, cosmopolitan, and consumerist." Our argument is that they are often symbiotic and facilitate each other.

However, if the narrative of the citizen worker during much of the twentieth century was *inclusion* in the fold of the state, the same is not true of the contemporary era. Instead, the logic of the global citizen is *exclusivity*: she or he is groomed, sorted, and selected. Accordingly, we start this discussion by noting that the creation of a global-citizen workforce is a deeply contentious process, which is itself rooted in the internal complexities of globalization.

A helpful framing is the neoliberal exercise of "graduated sovereignty" by state governments (Ong, 2006). The formation of special economic zones, which have proliferated since the 1960s in many cities around the world, is an example. These zones often differentially affect citizens within and without their boundaries. Located in countries such as Taiwan, China, the Philippines, Mexico, and the Dominican Republic, they have served as a strategy through which states can attract foreign capital, for instance, by providing tax havens and corporate incentives to multinational firms. Often, however, their political and governance structures are enhanced in relation to the nation within which they exist (Ong, 2006, p. 104). As a result of yielding to global capital, state control over peoples within the physical boundaries of the nation is heightened. Thus, rather than uniformly exercising law and policy on its citizens, states make different "biopolitical investments" in segments of the population, especially when it comes to the distribution of care, services, and protection (Nah, 2012).

One can witness this graduated sovereignty in action through the emergence of the global call centre industry. Just as nations are constructed as global and market driven, citizen workers are evaluated according to those qualities too. Tracing the promotion of the call centre sector in India's economic "miracle" (that is, its liberalization project), Rowe, Malhotra, and Pérez (2013) note a shift in focus from the *number* of Indian citizens to the *types* of Indian citizens: "Indian citizen-subjects are valued according to their capacity to participate in the global market; to be a global subject is to be a good Indian" (p.106). Indian call centre workers are made to imagine themselves in terms of the global utility of their labour, and it is this utility that determines who should belong to the "new India."

This volume reveals *call centres are sites in which the naming and unnaming of ideal global citizens occurs as part of the day-to-day work*. Take the case of the Philippines. Here, overseas foreign workers (OFWs) are state sanctioned as national assets. Along these lines, call centre workers occupy a similar ideological space: they are seen as good, responsible citizens who are national heroes because of their contribution to the local economy (Salonga, chapter 5). This is not coincidental. Indeed, financial outputs of these two industries run in parallel; reports indicate that the customer service sector now accounts for 10 per cent of the Filipino economy, thus generating the same revenue as that of remittances from eleven million overseas Filipinos (Lee, 2015). Return on investment, therefore, can be a critical factor in inclusion as a valued labourer for the nation. Our cases show how other qualities of the labour force play into this process; class and educational status also have a bearing on the selection of the global citizen worker. In Guyana (chapter 3), like the Philippines, call centres are populated by middle-class, educated workers whose social positions allow for relatively easy connections to national heroism.

Not all call centre workers can be converted into celebrated national heroes, however. Foretelling the significance of this issue, Mehta-Karia (2011) asks, "What does the exaltation of the branded citizen mean for the existence and welfare of the unbranded subject?" (p. 9). Our authors have many insightful responses, especially about the complicated cases – workers who are not automatically welcomed as global citizens, even though in practice they are doing such labour for call centres. For instance, in chapters 2 and 6, respectively, we find that call centres in El Salvador and Guatemala are recruiting employees who have been deported from the United States. This reflects a search for perfectly bilingual workers who can embody the national brand. Yet these workers pose a significant challenge to the ideological construction of the national hero. They may possess the exact linguistic and cultural skills needed for the industry, but, significantly, they are often associated with criminality.

The "deportee call centre worker" is emerging as an important figure in studies of other call centre settings. Anderson's (2013) study of

Mexican call centres reveals the ways in which the transnational call centre sector plays a unique role in forging the link between punitive immigration policy and global capitalism. She traces the proliferation of transnational call centres in Mexico between 2000 and 2010. In her case study of one organization employing 44,000 workers, Anderson found that 30 per cent of the workforce had in fact been deported from the United States. In 2011 the United States repatriated more than 350,000 Mexican migrants (Flannery, 2014). Anderson notes, "English language and US American fluency became a marketable skill for the returning migrant who regularly talks to people within the United States and yet cannot legally set foot in the country he or she calls home" (2013, p. 3). Callers commend such workers on their English-language skills, and, for deportees, working in a call centre holds "bittersweet moments of recognition and belonging" (2013, p. 5). The creation of the global Mexican subject does not just occur through an alliance between the Mexican state and transnational corporate investors. U.S. state deportation policy creates workforces within Mexico that are in turn hired (for considerably lower wages) by transnational firms to serve American customers.

Transnational Communication Hierarchies

Call centres are settings in which workers and customers do not come into physical contact with one another but are instead "voice embodied" (Mirchandani, 2015). These voice-embodied workers are the new service providers globally, as the work of customer service has shifted dramatically in recent decades. It has transformed from localized, individualized interactions to remote outsourced, global interactions. A central dimension in voice-based customer service work, however, is that it occurs through language. In the logic of the call centre industry, linguistic commonalities have the capacity to forge connections between nations and between the workers and customers within them.

The strategy of the global call centre industry for achieving this linguistic symmetry is to draw upon the geographies of the past. Ironically, within a sector said to represent the forward-looking vision of technological capability, call centre personnel have recast the nation's colonial history (and its accompanying educational and linguistic infrastructures) as transnational assets. Playing out the post-colonial moment, the industry emphasizes how global south countries can "reconnect" with their prior colonizers, now as waged service providers. In some cases, this is not just a dyad but a multiplicative set of colonial relations. For instance, Mauritius (chapter 4) is seen as an ideal location from which both French- and English-speaking customers can be served, given its double colonization by France and then Britain. Guatemala (chapter 6) and El Salvador (chapter 2) attract transnational call centre companies on the basis of their bilingual, Spanish-English-speaking subjects. Thus, numerous communicative symmetries can sometimes be forged within a single national setting.

Yet, this ideal of creating commonalities through language inevitably encounters barriers. The underlying complications of the post-colonial period tend to be reproduced in the form of a *global communication hierarchy*. As reflective of power imbalances in the global economy, some countries are ranked and valued above others in terms of their languages and ways of speaking. This is reflected in a resurgence of nationalism within the service industry. In call centres both customers and workers are found to express national allegiances (and, alternatively, ethnocentrisms) via information and communication technologies.

This dynamic surfaces within global call centres in the form of accent stereotyping. Research shows that customers (across service industries) frequently engage in discrimination concerning speech (Hill & Tombs, 2011). For instance, U.S. or French accents are associated with sophistication, while African American or Mexican American accents are associated with low socio-economic status. In a study of Australian customers, Hill and Tombs find that people attribute negative traits to accents diverting from standard Australian English. Moreover, they evaluate not only *the employees* with those accents but *the service* they provide, as inferior. Customers have less confidence in the communication skills of workers who have accents different from their own (with the exception of certain business enclaves, such as international restaurants, where servers' accents signify the authenticity of the cuisine). The same is true of call centres, where customers (especially when they are already frustrated by the problem about which they are calling) often stereotype employees with accents from previously colonized countries as poor communicators.

In their everyday activities on the telephone, workers are obliged to cross these linguistic hierarchies. Accordingly, managing communications and the tensions therein becomes one of the central tasks for global call centres. Many firms have responded by using language training as way to reset linguistic and cultural commonalities between global north customers and global south workers. In India, researchers have documented the near universal use of such programs. For some workers the process starts even before they arrive at the call centre. A largely unregulated but burgeoning industry of language training has arisen in India, convincing prospective candidates of the need to undergo such programs prior to, during, and/or between jobs. Once employed, most new recruits also spend up to three months learning what are termed "neutral" accents (that is, those that customers can understand). The focus is on the removal of "mother-tongue influences," which are deemed localized and therefore inappropriate for global subjects.

Studies in this volume show how such training is not limited to India and is in fact global. It is prevalent in the Philippines, as Salonga (chapter 5) reports. Here employees are trained to speak in English accents that will be understood not just by Americans but by English speakers in all customer countries. This type of accent training feeds into the broader cosmopolitan identity that workers are expected to develop. Salonga argues that, rather than being "placeless," such linguistic cosmopolitanism is closely tied to the nation. Specifically, it reflects assumptions about the "natural" propensity of Filipinos to speak in sweet, caring, and nurturing ways. As such, it represents another example of accent stereotyping – yet, in this case, from inside the firm and nation. It emerges not only transnationally from customers but also locally from state officials, company executives, and even other workers.

Some countries in our study, interestingly, do not engage this strategy of accent training – but with consequences. Guyana's linguistic and cultural closeness to the United States, framed as the ideal quality for the outsourcing of call centre work, is a focal point of state branding practices (chapter 3). As such, potential employees are often screened for the correct linguistic traits. In the context of severe cost-cutting pressure, furthermore, there is little basis for workers or state officials to advocate for more resources. In turn, accent training is not widespread in Guyanese call centres. Nonetheless, like Indian and Filipino workers, employees in Guyana report extensive customer anger when accents are deemed inappropriate. Thus, global communication hierarchies surface and remain apparent and visible in the day-to-day conversations on the telephone.

The impact of linguistic hierarchies is both subtle and overt. More nuanced is the way in which such practices obscure the underlying foundations of neoliberalism in these locations. Narratives and constructions of language in global call centres operate by taking advantage of (while also suppressing) the colonial histories upon which they are based.

Just like the case of India, where assumptions of "neutral" English mask legacies of British occupation and "civilizing" practices of speech, "neutral" Spanish in El Salvador entrenches further the "vanishing" of the country's indigenous past, as Rivas (chapter 2) notes.

In addition, linguistic hierarchies feed into the project of making and unmaking the global citizen worker, as discussed earlier. Accent acquisition and training become primary tools for defining national subjects as global versus non-global citizens. For instance, some Indians are considered to be more trainable within call centres – those who are middle class and educated at formerly British missionary institutions, like Catholic convents and Jesuit schools, which are largely English medium. In contrast, those who were educated under regional languages and are therefore multilingual are deemed likely to "corrupt" the English pronunciation. A similar trend occurs in El Salvador. Differentiations and stratifications are made among various groups who speak "perfect English." As Rivas's ethnography reveals (chapter 2), the way in which an employee acquired that skill matters – whether by virtue of an upper-class, locally based education or by residence in (and then deportation from) the United States. In the latter case, potential applicants with perfect English are not valued but instead are suspected of criminal and gang-related tendencies. Ironically, then, perfect English can be a marker of non-worthiness as a global citizen worker.

A third impact of linguistic hierarchies is the structuring and stratification of rewards. This is evident, for instance, in the valuing or devaluing of multilingualism as an occupational skill. Our studies find that language expertise is remunerated in vastly different ways across contexts. In Morocco (chapter 7) some of the call centre workers who were interviewed earn close to the wages paid to engineers, as they seamlessly translate standardized scripts in English, French, German, and Spanish in the course of their interactions. In contrast, in El Paso (chapter 8), bilingualism is seen as a product of workers' border location, rather than as a payable asset. Therefore, despite the linguistic requirements of the job, employees are paid lower than minimum wage. Spanish speakers within border communities in Texas are assumed to have limited English proficiency, and this in turn justifies their poor wages.

These stratifying mechanisms of linguistic resources serve as a poignant reminder of the socio-political construction of skill by nation. Entrepreneurs in the call centre industry frequently promise the generic and transferable nature of skills in their outsourced labour processes, which in turn fuels the development of global education and training industries. Yet the actual determination of skills and their economic value depends on the context. Differential social constructions of language in the workplace are stark reminders of the limitations of human capital approaches. Chapters in this collection suggest that, rather than enhancing employment potential, language expertise can at times *negatively* "mark" individuals – as national outsiders, gang members, or criminals. The same trait can be simultaneously viewed in opposing ways depending on time and place; in particular, the capacity to speak a language fluently can be viewed either as "human capital" (a positively viewed skill) or "baggage" (a negatively viewed skill).

These enduring patterns of global linguistic hierarchy underscore the ways in which transnational call centre workers engage in communicative labour. Such work is structurally embedded in what Dean (2009) calls "communicative capitalism": "a political-economic formation in which there is talk without response" (p. 24). Network technologies, such as the high-speed cable networks through which transnational customer service exchanges occur, give rise to fantasies of abundance, participation, and wholeness. The multiplicity of interaction mediums provides hope for a situation in which all individuals can participate beyond their local settings. Such democratic assumptions underlying the proliferation of communications, however, are misguided. Dean notes that communicative capitalism in fact "strengthens the grip of neoliberalism" (2009, p. 49).

In call centres workers cross discursive boundaries between nations on a daily basis, and these crossings are constitutive of their identities as global subjects participating in a global world. Yet, their "participation" in transnational exchanges occurs in a pre-established hierarchy in which they have a voice only as servers. *Rather than communication (which assumes the understanding of messages between sender and receiver), call centre exchanges often facilitate the circulation of servitude and privilege* (Brophy, 2010). Dean (2009) notes that, under communicative capitalism, commercial choices are elevated as an expression of political choice. In line with this, customers protect their national assets (outsourced jobs) through expressions of anger towards service providers.

In these ways language and its embeddedness in social stratifications are integral to the dynamics of global labour in call centres. On a continuous basis, employees manage and cross linguistic hierarchies during telephone interactions between agents and customers.

Tensions of Servicing the Labour-Nation Nexus

On one level, the proliferation of transnational call centres – in which workers from one country provide live, voice-based, interactive services to those from another – is itself a powerful symbol of a world without borders. In the course of the provision and the consumption of services, citizens from different countries interact without checkpoints, passports, or fences. Indeed, states compete to attract global call centres, and nations without borders are strategically constructed as ideal locations for transnational call centres.

On another level, however, as the chapters in this collection attest, borders are continually enacted through and within call centres. For the agents of that process – workers who participate in the "border as method" (Mezzadra & Neilson, 2013), constructing national lines and boundaries through their transnational service work – the border zone can be advantageous. Workers benefit from its digital permeability in many ways. For instance, employees in Moroccan call centres (chapter 7) value the ability to travel internationally through the telephone, even when they are denied visas in the legal system of the material world. In other ways too, many workers are happy with their employment in global call centres. Having a job at all in economies that are struggling, not to mention a job in an office (rather than, say, in manual services), and one that is well paying (in some cases), is appreciated by many employees. Indeed, the positive features of call centre employment are often well circulated in the media and labour market by state offices and their advertising strategists.

At the same time, ethnographies in this book reveal that the nexus of labour and nation leads to a variety of contradictions and tensions, the brunt of which are experienced by workers. Several of the contributors lay out the contradictions that emerge between the borderless flow of global firms and the bounded citizens that comprise their labour force. Trotz, Mirchandani, and Khan (chapter 3) and Heyman and Alarcón (chapter 8) give vivid witness to the degradation of call centre employment and its impact on workers. Neoliberal economic programs and lean service production regimes have led to an expansion of jobs that have little stability, low wages, and few benefits. Employees in Guyana are at times remunerated in non-wages – grocery vouchers or "fun" work environments. Indeed, rather than the premium jobs as advertised and touted by state and industry leaders, call centre jobs can be dismal and prohibitive of a living wage. In some locations, like Morocco, the call centre "offers good money, but ... no future."

Aspects of the global call centre experience are discriminatory, hierarchical, and punitive. Guyanese call centre managers engage in school-like disciplinary methods (chapter 3); those in Guatemala (chapter 6) subject their workers to polygraph tests and strict surveillance. Even in Morocco, where earnings are reported to be higher than in other locations, workers lament the routinized nature of their work and report sentiments of alienation and imprisonment.

Furthermore, many of the chapters recount hierarchies in job ladders and rewards based on language, race, and background. On the U.S.-Mexican border (chapter 8) promotions from telephone worker to manager, and from branch operation to corporate offices, are awarded to workers with English accents. Spanish accents, as well as English-Spanish bilingualism, are considered unworthy of extra compensation, despite their assistance in serving multiple kinds of consumer populations.

The tensions of labour and nation manifest in other, more subjective, ways for employees. There are many forms of social disgrace and outcasting that workers face in these call centres. Contributors to this book discuss nationalized types of stigmatization in the workplace: of language (being a border English speaker), of territory (residing in a low-status part of the city), and of deportation (marked with a criminal status). Typically, "cosmopolitan" displays in the call centre are favoured over markers of "indigenousness."

Critically, it is often the workers who face the crucial role of mediating national and global identities in the call centre. The tasks of constructing a fictitious national location, or deceiving consumers, become part of the call centre routine for employees (Poster, 2007, 2013a). Workers face the personal turmoil of dispossession from one state or another, in the repertoire of nationalities with which they interact. In Morocco (chapter 7) workers talk about "becoming an alien in your own land" as they make daily journeys to Europe on the telephone but are "restrained from speaking Arabic" to each other.

Moreover, call centres compress and overlay meanings of nationhood that are directly in opposition to each other: citizen/non-citizen, alien/native, legal/illegal. Employers and their policies selectively deploy these concepts to fit their needs at particular moments. Take for example the deported "homies" in Meoño Artiga's analysis (chapter 6). Their "cholo" clothes, which were once outlawed for Chicanas in the United States, are accepted half a century later in Guatemalan call centres. Homies have traits that are valued by the firms (for example, telemarketing aggressiveness), more so than those of "local" workers. Yet they are also criminalized in other situations and treated as potential threats to the call centre. Deported workers may face a double criminalization – by state governments and then by call centre firms and their co-workers. For U.S. firms that subcontract their services to outsourced call centres, there is a profound, underlying contradiction: these U.S. companies are indirectly hiring employees who are considered to be criminals by their state government. This layering of mixed messages creates a highly contested work experience for employees.

Appadurai (2006) has noted that the creation of communities, of which nations are examples, entails violence. He argues that with the growth of globalization, including the expanded focus on free trade, transnationalization of work, and universal human rights, there has been a corresponding rise in ethnic cleansing and political violence. Those in power experience a "surplus of rage" against social or economic minorities, even those who may hold little power. Indeed, the chapters in this book show that call centre sectors are in fact sites within which inclusion and exclusion are enacted on a continuous basis. This is the interface between labour and nation – nations are created within organizational settings through the day-to-day work of employees, while citizens are deemed to be appropriate or inappropriate for employment in the nation's (often valorized) transnational industries.

Indeed, peppered throughout the studies in this book are cogent examples of the exclusionary ethnic stereotypes used to promote nations and citizens as attractive for transnational call centres. Many of these stereotypes are constructed as "aesthetic skills" that citizens possess by virtue of their nationality but also require training to hone and refine. Filipinos are national heroes, defined as sacrificial, sweet-sounding caretakers. Guyanese are constructed as cheap workers; they have high levels of education and can be easily understood by customers, but also hired for a fraction of the cost of labour in the United States or Europe. Guatemalan and El Salvadoran return migrants are reluctantly incorporated into the industry; they are proficient in English but also suspected of being gang members or criminals. Arabs are deemed appropriate call centre workers in Morocco because they are known for their trading, interaction, and negotiation skills. Mauritians are seen as more appropriate call centre workers than those with North African accents. Even within the United States, workers are stratified in their wages based on proximity to the borderland of El Paso. As Heyman and Alarcón note, "being a border Hispanic is a sign of exploitability." The call centre sector provides a clear illustration of the impacts of ethnic and national stratification: how it is made and remade on a daily basis, how it inevitably benefits transnational firms, and how it depresses worker power and wages.

Worker Agency and Resisting the Nationalist Paradigm

Workers play an active role in the exercise of nationhood within the call centre sector. As the interviews across numerous countries in this book show, workers are deeply aware of both their roles as citizens of nations and the impact of globalization. These cases illustrate the meaningful ways in which employees exert agency and subvert the manipulations of nationhood and national identity. Moroccan workers (chapter 7) use the global call centre to make "sneaky" discursive journeys to the global north when they are denied physical journeys owing to visa restrictions. Returned migrants in Guatemala (chapter 6) construct malleable identities to become "telephone gangsters," who channel their "homie" aggressions into making sales and become "decent," responsible employees, thereby surpassing the stigmatization of deportation. These examples reveal the subtleties of resistance in the local and global context.

In many instances workers also engage in direct protest. Employees in the Philippines (chapter 5) have taken to the streets to defend their livelihoods against the derogatory images of call centre workers on television. Employees in Guyana (chapter 3) go on strike to contest abuses of work time.

There are also examples of emergent transnational solidarity in these analyses. Heyman and Alarcón (chapter 8) note that language itself becomes a force of cohesion across the U.S.-Mexican border. By using informal dialogue, by code switching between languages, and by integrating Spanish and English, U.S. workers form bonds with Mexican American customers which in turn mute nationalism. While we found little direct evidence of worker alliances across outsourced sites, there is a clear overlap in concerns that suggests fruitful possibilities for transnational worker advocacy. Along these lines, Poster (2011) has reported active cross-border organizing by call centre worker associations in the United States and India, indicating that such trends are occurring. Nationalism can play a helpful role here as well. Scholars argue that, even in the era of "new labour internationalism," worker movements at the national level will be key to the momentum of future organizing across borders (Evans, 2014).

In the final analysis, this book illustrates that labour studies benefit from recognizing how nationhood is implicated, structured, and created through work. Similarly, explorations of nationhood are enriched by going beyond issues of state policy and border protection to discover the meanings of citizenship that arise in peoples' daily engagement with the labour market.

REFERENCES

Anderson, B. (1991). *Imagined Communities*. New York: Verso.
Anderson, J. (2013). The (re)productive age: Return migration, call centers, and the (re)production of culture. Seminario Internacional sobre Migración de Retorno, Mexico. http://www.academia.edu/3419871/_The_Re_Productive_Age_Return_Migration_Call_Centers_and_the_Re_Production_of_Culture_
Anholt, S. (2009). The media and national image. *Place Branding and Public Diplomacy*, 5(3), 169–79. http://dx.doi.org/10.1057/pb.2009.11
Appadurai, A. (2006). *The fear of small numbers: An essay on the geography of anger*. Durham, NC: Duke University Press. http://dx.doi.org/10.1215/9780822387541
Bajaj, V. (2011, 26 November). A new capital of call centers. *New York Times*. Retrieved from http://www.nytimes.com/2011/11/26/business/philippines-overtakes-india-as-hub-of-call-centers.html?pagewanted=all&_r=0
Billig, M. (1995). *Banal nationalism*. London: Sage Publications.
Blauner, R. (1972). *Racial oppression in America*. New York: HarperCollins.
Bonds, A. (2006). Calling on femininity? Gender, call centers, and restructuring in the rural American West. *ACME: An International E-Journal for Critical Geographies*, 5, 28–49.
Brophy, E. (2010). The subterranean stream: Communicative capitalism and call center labour. *Ephemera: Theory and Politics in Organizations*, 10(4), 470–83.
Dean, J. (2009). *Democracy and other neoliberal fantasies: Communicative capitalism and left politics*. Durham, NC: Duke University Press. http://dx.doi.org/10.1215/9780822390923
Evans, P. (2014). *National labor movements and transnational connections: Global labor's evolving architecture under neoliberalism*. UC Berkeley Institute for Research on Labor and Employment, Working Paper no. 116-14. http://irle.berkeley.edu/workingpapers/116-14.pdf.
Fan, Y. (2010). Branding the nation: Towards a better understanding. *Place Branding and Public Diplomacy*, 6(2), 97–103. http://dx.doi.org/10.1057/pb.2010.16

Flannery, N.P. (2014). Dispatches: Mexico's new demographic dividend. *Americas Quarterly*, Higher Education and Competitiveness issue (Summer).

Fox, J.E., & Miller-Idriss, C. (2008). The "here and now" of everyday nationhood. *Ethnicities, 8*(4), 573–6. http://dx.doi.org/10.1177/14687968080080040103

Graham, M., & Mann, L. (2013). Imagining a silicon savannah? Technological and conceptual connectivity in Kenya's BPO and software development sectors. *Electronic Journal of Information Systems in Developing Countries, 56*(2), 1–19.

Hill, S.R., & Tombs, A. (2011). The effect of accent of service employee on customer service evaluation, *Managing Service Quality, 21*(6), 649–66.

Hoenig, L.J. (2012). The branding of African American slaves. *Archives of Dermatology, 148*(2), 271. http://dx.doi.org/10.1001/archdermatol.2011.2683

Hossfeld, K.J. (1990). Their logic against them: Contradictions in sex, race, and class in Silicon Valley. In K. Ward (Ed.), *Women workers and global restructuring* (pp. 149–78). Ithaca, NY: ILR Press.

Kotler, P., & Gertner, D. (2002). Country as brand, product, and beyond: A place marketing and brand management perspective. *Journal of Brand Management, 9*(4), 249–61. http://dx.doi.org/10.1057/palgrave.bm.2540076

Lee, D. (2015, 1 February). The Philippines has become the call-center capital of the world. *Los Angeles Times*.

Mann, L., Graham, M., & Friederici, N. (2014). *The Internet and business process outsourcing in East Africa*. Oxford: Oxford Internet Institute.

Mehta-Karia, S. (2011) Imagining India: The nation as a brand. (Master's thesis). Department of Sociology and Equity Studies in Education, University of Toronto. Retrieved 5 December 2011.

Mezzadra, S., & Neilson, B. (2013). *Border as method, or the multiplication of labor*. Durham, NC: Duke University Press. http://dx.doi.org/10.1215/9780822377542

Mirchandani, K. (2012). *Phone clones*. Ithaca, NY: Cornell University Press.

Mirchandani, K. (2015). Flesh in voice: The no-touch embodiment of transnational customer service workers. *Organization, 22*(6), 909–23. http://dx.doi.org/10.1177/1350508414527779

Nah, A.M. (2012). Globalisation, sovereignty, and immigration control: The hierarchy of rights for migrant workers in Malaysia. *Asian Journal of Social Science, 40*(4), 486–508. http://dx.doi.org/10.1163/15685314-12341244

Nakamura, L. (2014). Indigenous circuits: Navajo women and the racialization of early electronic manufacture. *American Quarterly, 66*(4), 919–41. http://dx.doi.org/10.1353/aq.2014.0070

Noronha, E., & D'Cruz, P. (2009). *Employee identity in Indian call centres: The notion of professionalism*. New Delhi: Sage/Response.

Ong, A. (2006). *Neoliberalism as exception: Mutations in citizenship and sovereignty.* Durham, NC: Duke University Press. http://dx.doi.org/10.1215/9780822387879

Poster, W.R. (2007). Who's on the line? Indian call center agents pose as Americans for U.S.-outsourced firms. *Industrial Relations, 46*(2), 271–304. http://dx.doi.org/10.1111/j.1468-232X.2007.00468.x

Poster, W.R. (2011). Emotion detectors, answering machines and e-unions: Multisurveillances in the global interactive services industry. *American Behavioral Scientist, 55*(7), 868–901. http://dx.doi.org/10.1177/0002764211407833

Poster, W.R. (2013a). Hidden sides of the credit economy: Emotions, outsourcing, and Indian call centers. *International Journal of Comparative Sociology, 54*(3), 205–27. http://dx.doi.org/10.1177/0020715213501823

Poster, W.R. (2013b). Subversions of techno-masculinity in the global economy: Multi-level challenges by Indian professionals to US ICT hegemony. In J. Hearn, M. Blagojevic, & K. Harrison (Eds.), *Rethinking transnational men* (pp. 113–33). London: Routledge.

Qiu, J.L. (2010). Network labour and non-elite knowledge workers in China. *Work Organisation, Labour & Globalisation, 4*(2), 80–95. http://dx.doi.org/10.13169/workorgalaboglob.4.2.0080

Rowe, A.C., Malhotra, S., & Pérez, K. (2013). *Answer the call: Virtual migration in Indian call centers.* Minneapolis, MN: University of Minnesota Press. http://dx.doi.org/10.5749/minnesota/9780816689385.001.0001

Sassen, S. (1996). *Losing control: Sovereignty in an age of globalization.* New York: Columbia University Press.

Wessler, S.F. (2014, 14 March). No place like home. *This American Life, 520.*

Contributors

Amado Alarcón is a professor titular in sociology (equivalent to an associate professor) at Rovira i Virgili University, Reus, Spain. His publications include about fifty scientific articles and five books devoted to language, migration, and labour markets. He has been commissioned as principal investigator by the Government of Catalonia (Secretariat for Immigration; Catalan Observatory on Youth) and the Government of Spain (Ministry of Education and Science and Ministry of Economy and Competitiveness). He is president of Research Committee 25 Language & Society, International Sociological Association, and was program coordinator for this committee at the ISA II Forum of Sociology (Buenos Aires, 2012) and the ISA XVIII World Congress (Yokohama, 2014).

Chris Benner holds the Dorothy E. Everett Chair in Global Information and Social Entrepreneurship, is director of the Everett Program for Digital Tools for Social Innovation, and is a professor of environmental studies and sociology at the University of California, Santa Cruz. His research examines the relationships between technological change, regional development, and the structure of economic opportunity, focusing on regional labour markets and the transformation of work and employment. His most recent (2015) co-authored book, *Equity, Growth, and Community: What the Nation Can Learn from America's Metro Regions* (University of California Press), examines the diversity and dynamics of regional knowledge communities, and their relationship to social equity and economic growth.

Sanae Elmoudden is an associate professor at St John's University, Boulder, Colorado, teaching classes in organizational, interpersonal, and intercultural communication. Her research and teaching address issues of globalization and communication technology discourses and their impact on cultural and organizational communication. Her main interests are in global organizing and offshore organizations in the MENA region, Morocco in particular. Elmoudden focuses specifically on the intersections of the local context, communication technology, and global spaces. Her theoretical objective is to explore new ways of conceptualizing place and crossing, helping the communication community to understand better the connections between the physical environment and the symbolic ones that shape us under globalization.

Josiah Heyman is a professor of anthropology and holds a chair of sociology and anthropology, University of Texas at El Paso. A long-standing scholar of the U.S.-Mexican border and border studies generally, he is the author of two books, the editor of a third book, and the author of over 120 scholarly articles, book chapters, and other scholarly writings. He has been funded by the National Institutes of Health and the National Science Foundation. Heyman also works closely with border human rights non-governmental organizations, in particular the Border Network for Human Rights in El Paso, Texas.

Iman Khan has been a freelance consultant in the field of women's and children's human rights and organizational development for the past three years. She was born in Guyana and educated in Canada with undergraduate degrees in Latin American and Caribbean studies and political science, and a master's degree in political science. Her most recent research in Guyana has revolved around the role of women in politics in Guyana and in project evaluation. Khan's work with women in the business community has also led her into the field of entrepreneurship with an emphasis on management, consultancy, and training in the private sector and with non-governmental organizations.

Luis Pedro Meoño Artiga has a bachelor's degree in anthropology from the Universidad San Carlos de Guatemala and a master's degree in anthropological sciences; he is currently a doctoral candidate at the Universidad Autónoma Metropolitana in Mexico. He has worked as an investigator for the Center of Regional Investigation in Mesoamerica and the Rigoberta Menchú Tum Foundation, examining the current state of inter-ethnic relations in Guatemala. His academic interests are focused on urban anthropology, in particular Guatemala City, from contemporary expressions of popular culture to return migration, and the new labour issues faced by workers in the transnational industry of call centres.

Kiran Mirchandani is a professor in the Adult Education and Community Development Program at the University of Toronto. Her research and teaching focuses on gendered and racialized processes in the workplace; critical perspectives on organizational development and learning; criminalization and welfare policy; and globalization and economic restructuring. Mirchandani is the author of *Phone Clones: Transnational Service Work in the Global Economy* (2012), co-author of *Criminalizing Race, Criminalizing Poverty: Welfare Fraud Enforcement in Canada* (2007), and co-editor of *The Future of Lifelong Learning and Work: Critical Perspectives* (2008).

Winifred R. Poster teaches at Washington University, St Louis. Her interests lie in the rise of the global information and communication technology workforce, and the ways in which it affects women, ethnic groups, and low-income communities around the world. With South Asia as a focus, she has conducted ethnographies of outsourcing by U.S. firms to India over the past two decades. Current projects examine global circuits of software engineers and computer factory workers; transnational call centres and their practices of national identity management, reversal of work time, and multi-surveillances; and the gendering of cybersecurity, with its increasing role of women as leaders, designers, practitioners, and online spies. Her findings have appeared in journals such as the *American Sociological Review*; *Social Problems*; *Gender & Society*; *International Journal of Politics, Culture and Society*; *Journal of Developing Societies*; *Social Politics*; *Industrial Relations*; *Research in the Sociology of Work*; and *American Behavioural Scientist*. Her article "Who's on the Line?" has won awards from Emerald Management Reviews and the University of Minnesota.

Cecilia M. Rivas is an associate professor in the Department of Latin American and Latino Studies at the University of California, Santa Cruz. Her research and teaching areas include media and consumer cultures in El Salvador, nationalism, and contemporary Central America. Her first book, *Salvadoran Imaginaries: Mediated Identities and Cultures of Consumption*, explores a diverse range of sites in which national and transnational identity is imagined and forged. It was published by Rutgers University Press in 2014 as part of the series Latinidad: Transnational Cultures in the United States.

Jairus Rossi has a doctorate from the Department of Geography, University of Kentucky. He is interested in the way that land use patterns are shaped by citizen science initiatives, emerging technologies, and diverse cultural economies. His dissertation examined the use of genetics and seed banking in ecological restoration work. Rossi is a post-doctoral scholar with the Community and Economic Development Initiative of Kentucky.

Aileen O. Salonga is an associate professor in the Department of English and Comparative Literature in the University of the Philippines, Diliman. She has a doctorate in English language studies from the National University of Singapore, and a master's degree in English from Virginia Polytechnic Institute and State University. Her research areas include sociolinguistics, critical discourse analysis, gender and discourse studies, and translation studies. Currently her work revolves around the global spread of English and its politics within the context of globalization processes, specifically the offshore call centre industry in the Philippines.

Alissa Trotz is an associate professor in Women and Gender Studies and in the Caribbean Studies Program at New College, University of Toronto. She is also an associate faculty member at the Dame Nita Barrow Institute of Gender and Development Studies at the University of the West Indies, Cave Hill. Her research interests include the gendered politics of neoliberalism, social reproduction, and women's activism; gender, coloniality and violence; and transnational migration and diaspora. For the past seven years Trotz has edited a weekly newspaper column, "In the Diaspora," in *Stabroek News*, a Guyanese independent newspaper, and she is a member of Red Thread Women's Organization in Guyana.

www.ingramcontent.com/pod-product-compliance
Lightning Source LLC
Chambersburg PA
CBHW021854020426
42334CB00013B/325